High-Z Materials for X-ray Detection

Leonardo Abbene • Krzysztof (Kris) Iniewski
Editors

High-Z Materials for X-ray Detection

Material Properties and Characterization
Techniques

 Springer

Editors
Leonardo Abbene
Department of Physics and Chemistry
(DiFC) - Emilio Segrè
University of Palermo
Palermo, Italy

Krzysztof (Kris) Iniewski
Redlen Technologies
Saanichton, BC, Canada

ISBN 978-3-031-20957-4   ISBN 978-3-031-20955-0   (eBook)
https://doi.org/10.1007/978-3-031-20955-0

This Springer imprint is published by the registered company Springer Nature Switzerland AG
The registered company address is: Gewerbestrasse 11, 6330 Cham, Switzerland

Preface

Silicon (Si) and germanium (Ge) semiconductors played a key role in X-ray and gamma ray detection, with successful results in several application areas, from synchrotron science, homeland security, medical imaging, and astrophysics to non-destructive testing in food industry. Over the last two decades, several compound semiconductors with high atomic number (Z) and wide band-gap have been experienced, with the goal to increase the energy range and ensure room temperature operation in X-ray and gamma-ray detectors. Binary (e.g., CdTe, GaAs, HgI_2, TlBr), ternary (e.g., CdZnTe, CdMnTe), and quaternary (e.g., CdZnTeSe, In-GaAsP, InGaAlP) materials have been proposed, and continuous efforts have been made in the advancement of the crystal growth and device fabrication technologies.

In this framework, we are glad to edit this book addressing to the current state of the art in the development of high-Z materials for X-ray and gamma-ray detectors, with particular attention to the medical and astrophysical applications. The material properties of cadmium telluride (CdTe), cadmium zinc telluride (CdZnTe), and the quaternary CdZnTeSe detectors will be presented, focusing on the charge-carrier-transport properties and electrical contact characteristics. The charge sharing effects and original correction techniques in pixel detectors are also shown.

Palermo, Italy   Leonardo Abbene
Vancouver, BC, Canada   Krzysztof (Kris) Iniewski
2022

Contents

About the Editors

Leonardo Abbene is Associate Professor of Applied Physics in the Department of Physics and Chemistry Emilio Segrè at the University of Palermo, Italy. His research activities are mainly focused on the development and characterization of room temperature semiconductor radiation detectors (CdTe, CZT) and digital electronics for X-ray and gamma-ray spectroscopy and imaging. He was the principal investigator (PI) of several research projects, where innovative instrumentation has been developed for medical (mammography, computed tomography, BNCT) and astrophysical (focal plane detectors for X-ray telescopes) applications. He is author or co-author of more than 80 research papers in international journals and conferences. He is a frequent invited speaker and has served as chairman for international conferences. Prof. Abbene is reviewer and editorial board member for several international journals. In his free time, Leonardo likes to play tennis and fish in the Mediterranean Sea of the beautiful Sicily.

Krzysztof (Kris) Iniewski is a director of detector architecture and applications at Redlen Technologies Inc., a detector company based in British Columbia, Canada, owned by Canon Inc in Japan. During his tenure at Redlen, he managed development of highly integrated CZT detector products in medical imaging and security applications. Prior to Redlen, Kris held various management and academic positions at PMC-Sierra, University of Alberta, SFU, UBC, and the University of Toronto.

Dr. Iniewski has published more than 150 research papers in international journals and conferences. He holds more than 25 international patents granted in the USA, Canada, France, Germany, and Japan. He has written and edited more than 75 books for Wiley, Cambridge University Press, Mc-Graw Hill, CRC Press, and Springer. He is a frequent invited speaker and has consulted for multiple organizations internationally.

High-Z Materials for X-Ray and Gamma Ray Detection in Medical Imaging

Krzysztof (Kris) Iniewski and Conny Hansson

1 Introduction

Doctor's ability to diagnose and treat disease is dictated by his or her ability to investigate and understand the inner workings of living systems. Imaging technologies have become a central part of both everyday clinical practice and medical investigations. Their applications range from clinical diagnosis to advanced studies of biological systems. Their relevance lay in many diverse areas in biology and medicine, such as diagnosis of cancer, assessment of the cardiovascular systems, or applications in neurosciences, to mention few examples.

Medical imaging represents an increasingly important component of modern medical practice, since early and accurate diagnosis can substantially influence patient treatment strategies and improve treatment outcome and, by doing so, decrease both mortality of many diseases and improve quality of life. It can also facilitate patient management issues and improve health-care delivery and the effectiveness of resource utilization. The most current example is in the use of computed tomography (CT) to diagnose damage in lungs as a result of COVID-19 infection.

Depending on the physics and effects that are being used to obtain necessary imaging information, different modalities measure and visualize different characteristics of the investigated tissues or organs [13]. Attenuation of electromagnetic radiation lies behind X-ray imaging and computed tomography (CT); sound wave transmission and reflection are used in ultrasound (US) imaging; and magnetic resonance imaging (MRI) uses magnetic field that caused changes in hydrogen

K. (Kris) Iniewski (✉)
Redlen Technologies, Saanichton, BC, Canada

C. Hansson
Stanford Linear Accelerator Center (SLAC), Menlo Park, CA, USA

© The Author(s), under exclusive license to Springer Nature Switzerland AG 2023
L. Abbene, K. (Kris) Iniewski (eds.), *High-Z Materials for X-ray Detection*,
https://doi.org/10.1007/978-3-031-20955-0_1

1

state of water molecules. The images created by most of these modalities display anatomy of the body; however some like nuclear medicine or functional MRI can represent body functions. The book focuses on high-Z material detectors for X-ray-and gamma ray-based imaging modalities as they both have common denominator: detection of number of photons, their energies, and their positions.

Most medical X-ray imaging is based on a measurement of the X-ray beam attenuation. The attenuation is different for different tissues and may be additionally modified by the presence of contrast agents (e.g., intravenously administered iodine) if needed. There are several X-ray modalities widely used in medicine. Depending on their application, they can utilize different energies of radiation (e.g., lower energy in mammography than in radiography) or a different method of image creation (radiography vs computed tomography), but the basic scheme is always kept more or less the same: the patient is placed between a radiation source and radiation detector. The information from the detector is used to create an image.

In other medical imaging modalities, usually referred to as nuclear medicine, there is no X-ray tube, and the radiation source is directly injected inside the patient in the form of a radioisotope. The general term "nuclear medicine" encompasses several different imaging techniques, ranging from simple planar and whole-body studies to positron emission tomography (PET) and single-photon emission computed tomography (SPECT). All these diagnostic techniques create images by measuring electromagnetic radiation emitted by the tracer molecules which have been labeled with radioactive isotope and introduced into a patient's body. Additionally, nuclear medicine includes internal radiotherapy procedures where radioactive high-dose injections are being used in cancer treatment.

2 X-Ray and Gamma Ray Basics

2.1 Ionization Radiation

Ionizing radiation is radiation that is able to ionize atoms when passing through matter. Ionizing an atom occurs when radiation with enough energy removes bound electrons from their orbits leaving the atom charged. The most common types of ionizing radiation are alpha particles, protons, beta particles, neutrons, gamma rays, and X-rays.

Alpha particles are among the most massive types of radiation. They are nuclei of helium and, therefore, have two protons and two neutrons, so they are positively charged. Alpha particles are emitted during the decay of unstable heavy atomic nuclei. As an example of application in medicine, alpha particles are used to kill cancerous cells in a process called theranostics. A proton is a particle that is identical with the nucleus of a hydrogen atom, and they are positively charged. Protons, due to their electric charge, can be accelerated to a high energy with a particle accelerator.

Highly energetic protons are also used to kill cancerous cells in a procedure called external proton beam therapy.

Beta particles are either electrons with negative charge or positrons which are positively charged and have a much lower mass than alpha particles and protons. The negative betas come from nuclei that have too many neutrons, while the positive betas are generated from nuclei having too many protons. The electron and positron annihilation process is used in the imaging modality called positron emission tomography (PET).

Neutrons are neutral particles emitted from the nucleus of an atom during nuclear reactions. They are rarely used in medical imaging. Gamma rays are electromagnetic radiation similar to radio waves or visible light, except that they have much higher energies (shorter wavelength). After a radioactive decay (alpha and beta), the new nucleus often has an excess of energy that is usually released by the emission of gamma rays. Gamma rays are transmitted as photons, and in many ways, they behave more like particles than as waves. Gamma rays are used in a medical imaging modality called single-photon emission tomography (SPECT).

Finally, X-rays are also a form of electromagnetic radiation, similar to gamma rays, but they originate differently. X-rays originate in the orbital electron field surrounding the nucleus. This process takes place in its simple way when an electron is knocked out from its orbit, leaving the atom in an unstable state. Another electron of a higher energy falls down, filling the place of the missing electron, and the energy difference between the two energies is released in the form of an X-ray photon. Additionally, electrons may lose energy in the form of X-rays when they quickly decelerate upon striking a material. This is called Bremsstrahlung (or translating from German breaking radiation). X-rays are frequently used in medical imaging in form of radiography or computed tomography (CT). This chapter deals with applications of X-rays and gamma rays in medical imaging. Let us first review ways of photon interaction with matter, followed by a more detailed explanation of X-ray generation.

2.2 X-Ray and Gamma Ray Interactions with Matter

Energetic photons as X-rays and gamma rays interact with matter mainly by four basic processes: elastic scattering, photoelectric absorption, Compton scattering, and pair production. Elastic scattering, also called coherent scattering, is a process in which the photon path is altered without any loss of energy. Photoelectric absorption is the process where the photon disappears after transferring all its energy to a photoelectron. Most photoelectric interactions occur in the K-shell of the atom because the density of the electron cloud is greater in this region. Compton scattering is a process in which some fraction of the photon energy is transferred to a free electron in the material. The path of the photon changes. Finally, pair production is a process in which the photon energy is converted into an electron-positron pair.

All physical processes of X-ray interaction with matter (photoelectric absorption, Compton scattering, coherent scatter, and pair production) are used in various medical imaging modalities. The focus of this chapter is primarily on photoelectric effect in high-Z sensors where the transfer of the X-ray photon energy results in a directly measurable electric signal.

Photoelectric absorption is, in most cases, the ideal process for detector operation. All of the energy of an incident photon is transferred to one of the orbital electrons of the atoms within the detector material – usually to a K-shell electron. This photoelectron will have a kinetic energy equal to that of the incident photon energy minus the atomic binding energy of the ejected electron. The atom with a missing inner-shell electron emits a characteristic X-ray photon after an outer-shell electron is transferred to the incomplete inner shell. The emitted radiation is known as fluorescence. These fluorescent X-rays are unique fingerprint of the kind of atom that produces the fluorescence. This X-photon may interact with another atom by means of one or more processes until it loses all its energy or escapes from the material. Instead of emitting an X-ray fluorescence photon, the atom may emit an Auger electron with a characteristic energy. Auger electrons are of very low energy and can only travel a short distance within the matter. The Auger electron can escape from the material only when the absorption of the incoming photon takes place very close to the surface. In a semiconductor material as well as in other materials, the photoelectrons will lose its kinetic energy via Coulomb interactions with the semiconductor electrons, creating many electron-hole pairs.

Compton scattering is a collision between an incident photon and an orbital electron. The result is the creation of a recoil electron and a scattered photon. The incident photon energy is divided between the recoil electron and the scattered photon dependent on the scattering angle. In the case of multiple scattering, the Compton scattered photon further interacts with the radiation detector, either by additional Compton scatterings or via the photoelectric process. In this way, all the energy of the incoming photon might end up in the detector. The physics of Compton scattering can become somewhat complex but is useful for the design of Compton cameras.

Finally, for photon energies greater than twice the rest mass energy of the electron (1.022 MeV) pair production may occur and might actually be the dominant type of interaction at energies above several MeV. In this type of interaction, the energy of the incident photon is converted to an electron-positron pair. Afterward the positron annihilates with an electron producing two 511 keV photons. In most imaging modalities, photons with this high an energy are not used, so this effect can frequently be neglected.

2.3 X-Ray Generation

A very common X-ray source is the X-ray tube, schematically shown in Fig. 1. The operation of this device starts with the thermionic emission of electrons from

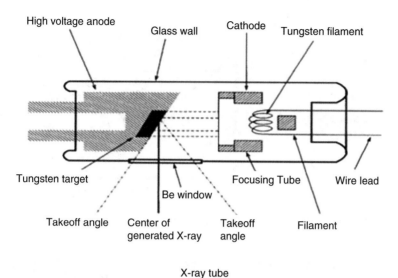

Fig. 1 Schematic representation of a typical X-ray tube

a filament. These electrons are subsequently accelerated by a voltage difference set up between the cathode and the anode (target). As the electrons slam into the target, they lose their gained kinetic energy to that target which results in the generation of X-rays.

X-rays produced in an X-ray tube are generated using two different mechanisms: Bremsstrahlung and the emission of characteristic radiation. The word Bremsstrahlung is retained from German language, roughly translatable to breaking radiation, and is used to describe the radiation which is emitted when electrons are decelerated in a material. It is characterized by a continuous distribution of X-ray intensity with energy and shifts toward higher energy when the energy of the bombarding electrons is increased. Characteristic X-ray on the other hand produces peaks of intensity at particular photon energies, the energy of which is set by the available atomic transitions of the target material, and these peaks are illustrated in Fig. 2. In practice, the emitted radiation is filtered by self-absorption within the target material, producing a high-pass filter response as low energy radiation is completely attenuated.

X-ray generation is an inefficient process as most of the kinetic energy of the emitted electrons are dispersed as heat in anode. Consequently, X-ray tubes generate a lot of heat, and heat extraction problems are primary concerns in the equipment design, manufacturing, and operation. In addition, only a few percent of the generated X-rays reach the detector as the X-ray beam is not generated directionally and photons are radiated in all directions. X-ray photon energy is related to acceleration voltage between the cathode and the anode, and if the acceleration voltage is 20 kV, the highest photon energy generated in the spectra

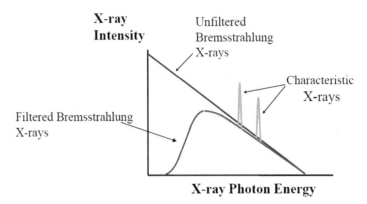

Fig. 2 Typical X-ray spectrum indicating Bremsstrahlung and characteristics X-rays

will be 20 keV. A typical X-ray system uses step-up transformers to produce the high voltage (HV) applied between the cathode and anode, and total number of photons generated is proportional to the cathode current, which typically is several milliamps (mAs).

X-ray tubes used in computed tomography (CT) are subjected to higher thermal loads than in any other diagnostic X-ray application. In early CT scanners, stationary anode X-ray tubes were used, since the long scan times meant that the instantaneous power level was low. Long scan times also allowed significant heat dissipation. Shorter scan times in later versions of CT scanners required high-power X-ray tubes, necessitating the use of liquid-cooled rotating anodes for efficient thermal dissipation. The recent introduction of helical CT with continuous scanner rotation placed even more demands on X-ray tubes which clearly is a challenging engineering problem as the dissipated power is in the kW range. A schematic representation of a CT scanner is shown in Fig. 3. The main components of the scanner are X-ray tube, detector array, and data acquisition system (DAS).

One of the concerns when using X-rays for imaging applications is the radiation damage they can cause. Since X-rays are ionizing radiation, at significant dose, they will cause tissue damage. The traditional unit of absorbed dose is rad. 1 rad is defined as an amount of X-ray radiation that imparts 100 ergs of energy per gram of tissue or re-stated in SI units as causing 0.01 Joule of energy to be absorbed per kilogram of matter. As a frame of reference, a typical chest X-ray exposure is about 50 mrads, while exposure of 50 rads causes radiation sickness. In the SI system, rad has now been superseded by gray with a simple relationship between the two as 1 gray equals to 0.01 rad.

In X-ray medical imaging, minimizing radiation dose is of critical importance. Equipment manufacturers have been minimizing the radiation dose for several years by employing the best detector technologies, optimizing image reconstruction techniques and selecting radiation dose-efficient clinical protocols. In that context,

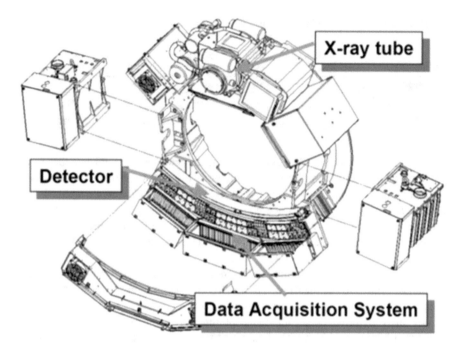

Fig. 3 Schematic drawing of a computed tomography (CT) scanner

development of the most efficient X-ray radiation detector that utilizes high-Z material continues to be very important.

2.4 Radiation Detectors

Operation of any radiation detector depends on the manner in which the radiation to be detected interacts with the sensor material of the detector. The interaction should cause a measurable change in a property of the sensor, typically a change in the number of free carriers available to conduct current. Electronic circuits then measure, and further process, this change in sensor property in order to extract information about the detected photon.

Radiation detection can be divided into two principal groups: indirect and direct detection. In the indirect process, scintillation materials usually perform detection, while for direct conversion, semiconductors are commonly used. Scintillators are materials that absorb photons of higher energy and re-emits photons of lower energy, usually in the visible part of the spectra. The visible light is subsequently detected by a photosensitive diode or similar opto-electronics device, which converts the light to an electrically measurable signal.

Table 1 List of some semiconductor materials used in radiation detection

Material	Germanium (Ge)	Silicon (Si)	Cadmium zinc telluride (CZT)	Thallium bromide (TlBr)	Mercuric iodine (HgI2)
Average Z	32	14	49	80	80
Density (g/cm^3)	5.32	2.33	5.78	7.56	6.4
Resistivity (Ohm-cm)	50	1E5	1E10	1E12	1E12

In the direct detection process, the semiconductor sensor converts the X-ray or gamma ray photon directly into a number of electron-hole pairs. The resulting current signal can then be directly measured with an electric circuit. Table 1 lists some semiconductor materials that have been used as X-ray sensor materials. With the exception of silicon and germanium, that are suitable for low-energy detection, all remaining materials, CdTe, CdZnTe, GaAs, TlBr, and HgI$_2$, are high-Z materials which are the subject of this book.

The design of the detector and the electronics allows one to extrapolate different properties related to the measured photon. The most common properties of interest are position of interaction, photon energy, and time of arrival of the photon. Position is obviously required to get an image of the source, by recording in detail the positions of the interactions of the incoming radiation. Energy is required to measure the spectrum of the source, i.e., a measurement of the energy deposited by each interacting photon. The time of arrival can be used to perform timing measurements and evaluate how long the photon has traveled w.r.t. a set time frame (time of flight).

3 Imagining Using X-Rays

3.1 Radiography

In the early days of X-ray imaging, glass plates with photographic emulsion were used as detectors, later replaced with films. The film is coated with a thin layer of photographic emulsion containing silver halide, which is not a very efficient X-ray absorber. The film performs much better in detecting visible light than X-rays. This is why the film has been quickly accompanied with an intensifying screen. The intensifying screen is a sheet of fluorescent material, converting the energy of each absorbed X-ray photon to many visible light photons. When the screen is in direct contact with the film, the film gets exposed by visible light. A combined screen-film detector has a much higher sensitivity for X-ray detection than the film alone. The sensitivity depends on the material used on the screen. As an example, screens with rare earth elements more effectively absorb and convert X-ray radiation compared to calcium tungstate screens (CaWO$_4$). With screen-film detectors, it is possible to obtain a radiographic image with a reasonably low patient dose. In medicine, the film is almost always used with intensifying screens, placed permanently in a light-tight radiographic cassette. In most cases, two screens are placed on both sides of the film, which has two layers of emulsion on both sides. The light emitted from the intensifying screen can penetrate the film base, reaching the second layer of emulsion and causing image blur. Compared with the film-only detector, the use of the screen-film decreases patient dose, but at the cost of a lower spatial resolution of the image. Similarly, a film with larger silver halide grains is more sensitive but has a lower spatial resolution. In mammography, a good spatial resolution is essential for good visualization of microcalcifications, and only one intensifying screen and one layer of emulsion are usually used. In dental radiography, small objects are imaged, and a good spatial resolution is important, while radiation dose is not of much concern, and the film can be used alone [30].

Radiographic films require photochemical processing to convert latent images to visible images. It is a time-consuming process, it is not environmentally friendly and has to be done in the darkroom. Various alternatives have been proposed over the years. Some of them gained initial interest but then were practically forgotten due to low performance, high cost, or impracticality. As an example, the xeroradiographic detector consists of a plate covered with an amorphous selenium photoconductor. Before use, the surface of the plate is uniformly positively charged. During exposure, electrons are excited in the photoconductor, causing local loss of positive charge. After exposure, a negatively charged toner is applied. The toner is attracted to the residual positive charge on the plate; then it can be moved to paper or plastic sheet, creating an image [33]. Interestingly, films that do not require photochemical processing and can be used in daylight conditions also exist. Radiochromic films become blackened directly by ionizing radiation, thanks to radiation-induced polymerization of diacetylene (dark as polymer, colorless as a monomer). Radiochromic films are not sensitive enough to be used for imaging, but they may be used as radiation detectors to monitor skin doses in interventional radiology.

Modern radiography has largely moved away from film detectors toward digital imaging. Computed radiography (CR) imaging plates and digital radiography (DR) flat-panel detectors do not need photochemical processing to get an image. They also have a much wider dynamic range than screen-film detector, e.g., 10^4:1 instead of 10:1 [8], and offer possibilities for digital processing, enhancement, and storage of the images.

In the CR systems, imaging plates are used as a radiation detector. The plate is placed in a radiographic cassette and emits visible light when absorbing X-ray radiation, which makes it similar to the intensifying screen in screen-film systems. At the same time, a sort of latent image is created on the plate, which makes it similar to film. During exposure, some electrons are trapped. During readout in the CR scanner, the imaging plate is illuminated with red laser, causing trapped electrons to relax to their ground state. The relaxation is associated with emission of blue light (photoluminescence), which is measured with a photomultiplier and used as an information to create an image. The light signal is proportional to the number of trapped electrons, which is in turn proportional to the received dose of radiation. Proper choice of storage phosphor materials for CR plates is essential for good performance of the system. The efficiency of radiation-to-light conversion of the material is important, but other characteristics may be important as well. Some materials (e.g., CsBr:Eu2+) perform better than others (e.g., BaFX:Eu2+) because of their oriented, needle-like structure, which resembles a matrix of optical fibers. Such a structure reduces light scatter in the readout phase, thus resulting in less image blur even if the material layer is made thicker to achieve higher sensitivity [8].

In screen-film systems, the energy of the X-ray radiation is converted to the image with a time-consuming photochemical process. In CR systems photochemical processing is not needed, but the plate is still a passive detector – it operates without any active means but does not provide a direct readout. Scanning of CR plates takes time; an image is available a few minutes after exposure. In DR flat panel detectors, the absorbed energy is converted to an electric charge within a detector, and the image is practically instantly displayed on the computer screen. In a vast majority of clinically used radiographic DR systems, the conversion is done with two steps, with intermediate conversion to light. First, X-ray radiation captured in a scintillation layer is converted to visible light. Second, the light is detected by a matrix of photodiodes. Electric charge is stored in a capacitor and then read out, amplified, digitized, and converted into an image. DR detectors with such indirect conversion are often referred to as aSi (amorphous silicon), although their amorphous silicon elements are not directly used in the detection of X-rays. This is done with the scintillation layer, which is made, e.g., of Gd_2O_2S or CsI. The materials differ in structure similarly as those used for CR imaging plates – while Gd_2O_2S is granular, CsI has a needle-like structure.

Direct-conversion DR systems also exist and are often used in mammography. They represent the latest improvement in radiation detection technology. The energy of the absorbed X-ray radiation is converted directly to electrical charge within the amorphous selenium (a-Se) or silicon detector materials. Electrons excited to the conduction band move along electrical field lines, without spreading to the sides, and are collected by the capacitors, similarly as in indirect DR detectors [28]. Higher-Z materials like GaAs, CdTe, and CdZnTe are used in direct conversion detectors for higher X-ray energies in the 40–180 keV range.

3.2   Fluoroscopy

The screen-film detector integrates signals during whole exposure, which makes it suitable to capture an image of a stationary object. If the object moves, its image is blurred. Movements of the patient during imaging can be minimized with short exposure time and with immobilization. Patients can be asked to hold a breath, but it is not possible to stop internal movements such as heartbeat or peristalsis. In some situations, it is essential to visualize movements within a patient. Besides the movement of the patient's tissues, the flow of the contrast medium may be observed, movement of surgical instruments, or movement of implants, which are being placed

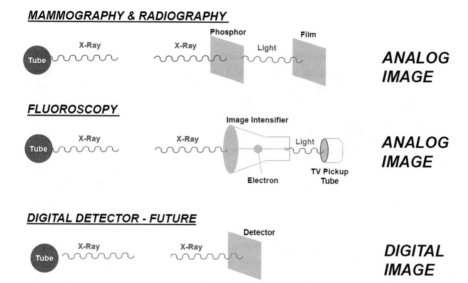

Fig. 4 Comparison between traditional X-ray modalities (mammography, radiography, fluoroscopy) and future digital detection that frequently utilizes high-Z detector materials

in the patient. That is why a radiation detector is needed, which can provide a live view with a good time resolution.

In the early days of radiology, the live image was observed directly on the fluorescent screen. This resulted in large radiation exposure not only for the patient but also for the operator, who often kept the screen in hand. Since the image was not very bright, fluoroscopy could be only used in darkened rooms, which was inconvenient. In the 1950s, a new image detector for fluoroscopy was introduced, namely, X-ray image intensifier. The operation of the image intensifier consists of several steps. X-ray photons interact with a scintillation layer, converting each absorbed X-ray photon into many visible light photons (as with a simple fluorescent screen). The light strikes photocathode, knocking out electrons, which are then accelerated with electric potential and then fall again on another (smaller) fluorescent screen. As an effect, there are many more photons on the smaller screen, resulting in a much brighter (intensified) image with a smaller dose to a patient. The bright image can be observed in normal lighting conditions, recorded with a video camera, etc. Currently, image intensifiers are being replaced with flat panel (DR) detectors with a fast readout [1]. High-Z materials will play a role in future evolution of the flat-panel detectors but require reduction in costs of these new detector technologies.

The future of many X-ray modalities relies on use of direct conversion detectors that enable digital detection. A schematic comparison between different X-ray modalities using indirect and direct detection schemes is shown in Fig. 4.

3.3 Tomographic Imaging

In radiography, three-dimensional structures are visualized in a two-dimensional planar image. The third dimension is lost in the imaging process. Images of objects located at different depths overlap, making some anatomical structures invisible. From the early days of medical X-ray imaging, researchers have tried to image selected planes within the patient. Many experiments involved movement of the radiation source (tube) and/or detector (film) during the exposure. If the tube and detector are synchronously translated in opposite directions during exposure, only the objects located at one plane will be projected all the time at the same positions on the detector, resulting in a sharp image. For other objects, their projected image on the detector will move, resulting in blur. This idea was used in classical tomography to obtain a sharp image of anatomical structures located at the focal plane. Typically, only the image of one plane was obtained in one exposure. Another film and other exposure had to be used to image structures located on another plane. It was not only time-consuming but also associated with a multiplication of patient exposure. This limitation can be overcome quite easily if a different detector is used. If the film is replaced with a flat panel DR detector, tens of images at different angles can be acquired during the movement of the tube-detector system. If the images are simply summed up together, the same image as in classical tomography is obtained. If the images are shifted before summation, a different plane is sharply imaged. Thanks to an image detector with good time resolution and a very simple shift-and-add algorithm, images of several planes can be obtained from a single exposure. This solution, called digital tomosynthesis, is used clinically [24], although usually in a slightly different scheme (X-ray tube is moving over a segment of a circle, while the detector remains stationary).

Digital tomosynthesis is currently used mainly in breast imaging. Cross-sectional images of head and body are usually obtained with X-ray computed tomography (CT). During the examination, tube and radiation detectors rotate around a patient, and a cross-sectional image is reconstructed with mathematical algorithms. Despite similarities between digital tomosynthesis and computed tomography, the geometry of the system is fundamentally different. In digital tomosynthesis, the image plane is perpendicular to the plane, in which the tube moves. In computed tomography, image lays in the tube rotation plane.

In current multislice CT systems, there may be several rows of detectors (e.g., 64 rows with ca. 1000 detectors in each row). Detectors may perform 2–3 revolutions around a patient per second, registering the signal several hundred times during each rotation. Detectors for modern CT systems need not only to have good sensitivity and high dynamic range but also to be very fast and capable of working in large g-force conditions. For many years, xenon detectors have been used in CT, due to their good time characteristics (fast decay of signal). However, the low density of xenon – even pressurized – does not make it a very efficient detector for X-rays. Currently, CT detectors are usually made of scintillators and photodiodes [29]. The scintillators are chosen to have a high light output, but also a low afterglow. The performance of the detectors, and thus also image quality, can be improved not only by choice of a better scintillator or more efficient photodiode but also by moving the analog-to-digital electronics as close as possible to the detectors, to minimize electronic noise [23].

3.4 Spectral Imaging

Attenuation coefficients for X-rays depend on the energy of photons. This dependence is different for different materials and tissues. This fact is used in bone densitometry (DXA, dual-energy X-ray absorptiometry), in which bone mineral density is calculated by comparing attenuation of radiation for two different energies. The typical radiographic image provides information about the attenuation of the radiation by the tissues, but it does not bring direct information about the density of the tissue. Identical attenuation can be observed, e.g., for the thicker bone of lower density, as well as for thinner bone of higher density. In bone densitometry, quantitative information on bone density is calculated from absorption data measured for two different energies of radiation. Similar approach can be used to obtain other information. In contrast-enhanced spectral mammography, two exposures are made for two different beam energies (different tube voltage and different filtration). Both exposures are made after administration of an intravenous contrast agent, but detector data obtained for two energies may be processed to create virtual non-contrast image [26]. Possibility of using two energies to determine electron density and effective atomic number of the examined tissues has already been proposed for the first commercially available computed tomography system. Currently, spectral CT allows for virtual monochromatic imaging, creation of virtual non-contrast images, better quantification of iodine content, and differentiation of renal stones based on their atomic composition [11, 18]. In some spectral computed tomography systems, the examination has to be performed twice to obtain data for two energies. Since the datasets are not obtained at the same time, patient movement can be an issue. Other spectral CT systems are capable of truly simultaneous imaging for two different energies, thanks to duplication of the tube-detectors system or fast alternating tube voltage switching [16].

All the spectral imaging methods mentioned so far are based on the use of two radiation beams with two different energies. Another approach is also possible, based on a single beam, and provided that detector separately measures signals in two (or more) energy ranges. In one design of a spectral CT scanner, a layered detector is used, with two layers sensitive to two different energy ranges [16]. Another possibility is the use of photon-counting detectors (PCD), e.g., cadmium telluride, or cadmium zinc telluride semiconductors [32]. The signal from most radiation detectors used in medical X-ray imaging is simply proportional to the total absorbed energy of radiation. In PCD, each detected X-ray photon generates a separately measured electrical pulse. The height of each pulse can be compared with the threshold, or several different thresholds, to assign it to one of the energy bins. For each pixel, PCD provides information on the number of pulses separately for each energy bin. This allows for truly simultaneous acquisition of separate images for several energy ranges during one exposure. If several contrast agents are administered to a patient before an examination, with different radiation absorption characteristics (e.g., iodine and gadolinium), the virtual reconstruction of images with individual contrast agents is possible. Besides energy-resolving capabilities, photon-counting detectors have no electronic noise, which allows better imaging [2].

4   Imaging Using Gamma Rays

4.1   Nuclear Medicine

In diagnostic X-ray imaging, the energy of X-ray radiation is optimized to achieve good visibility of anatomical structures with patient dose as low as possible. Depending on the application, tube voltage may be in the order of 25–45 keV in mammography, several dozen keV in radiography and fluoroscopy, and up to 120–140 keV in computed tomography (CT). This electric potential is used to accelerate electrons within the X-ray tube, and it sets a limit for the maximum energy of emitted photons (e.g., 140 keV in CT). Most of the photons reaching detectors have lower energy. However, imaging with photons of higher energy is also used in medicine. They are frequently emitted as gamma rays using radioactive sources, not from X-ray tube, and the corresponding imaging modalities are referred to as nuclear medicine.

Nuclear medicine (NM) imaging techniques, thanks to their inherently molecular characteristics and excellent sensitivity, have an important role to play, because they provide three-dimensional functional and in vivo information about biodistribution of tracer molecules labeled with radioactive isotopes [6, 7, 36]. For this reason, NM methods are often referred to as molecular imaging because they are able to localize molecular receptor sites, study biological markers expressed by diseased cells, and image the presence and extent of specific disease processes. The sensitivity of NM is several orders of magnitude better than MRI and CT for detection of metabolic changes in vivo. Additionally, NM offers an exceptional opportunity to combine diagnosis with treatment: When the same molecule is labeled with one isotope, it can be used to diagnose tumor; when it is used with another molecule, it can be used to carry chemo- or radiotherapy agents to treat the tumor; and, in parallel, it can be used to monitor effectiveness of this treatment. Monoclonal antibody and nanoparticle target molecule agents will provide such valuable alternative to traditional diagnostic and therapy procedures for oncology and cardiology.

The general term "nuclear medicine" encompasses several different imaging techniques, ranging from simple planar and whole-body studies to positron emission tomography (PET) and single-photon emission computed tomography (SPECT). All these diagnostic techniques create images by measuring electromagnetic radiation emitted by the tracer molecules which have been labeled with radioactive isotope and introduced into a patient's body. Additionally, nuclear medicine includes internal radiotherapy (IRT) procedures where radioactive high-dose injections are being used in cancer treatment.

In general, any nuclear medicine imaging study may be classified as (a) single-photon imaging, which measures photons emitted directly from a radioactive nucleus, and (b) coincidence imaging of pairs of 511-keV photons emitted in exactly opposite directions. These photons are created by positron annihilation following positron decay of the investigated nucleus. PET studies are acquired by cameras containing several small detectors arranged in multiple rings around the patient, while single-photon imaging studies use Anger cameras, with one, two, or even three large-area detectors.

For simplicity nuclear medicine imaging study may be divided into two classes:

- Single-photon imaging, which measures photons emitted directly from a radioactive nucleus, usually referred as SPECT (single-photon emission tomography)
- Coincidence imaging of pairs of 511-keV photons emitted in exactly opposite directions, usually referred as PET (positron emission tomography)

PET images are acquired by cameras containing several small detectors arranged in multiple rings around the patient, while single-photon imaging studies use Anger cameras, with multiple large-area detectors. New version of SPECT camera are being build that utilize Compton scatter and are therefore called Compton cameras.

4.2 Single-Photon Emission Tomography (SPECT)

In nuclear medicine imaging, a radioisotope is administered to the patient. After its uptake within the tissues, γ-radiation produced in the radioisotope's decay is detected from the outside of the patient. Tc-99m, which is the most widely used radioisotope for imaging, emits monochromatic γ quanta with an energy of 140 keV. Other isotopes used in SPECT are listed in Table 2, emitting gamma rays in a 70–440 keV range.

In static (planar) single-photon imaging, the camera is positioned next to the investigated organ, such as the heart or lungs, and the data are collected for several minutes. A two-dimensional map of radiotracer distribution is created. Planar images, however, suffer from poor contrast because photons coming from different layers in the body overlap and contributions from different organs cannot be separated.

Tomographic, three-dimensional (3D) images are created when series of two-dimensional (2D) planar images (projections) acquired at different angles around a patient are combined and reconstructed. These projections are acquired sequentially by rotating detectors in SPECT cameras.

Table 2 Radioisotopes used in SPECT imaging

Isotope	Energy (keV)	Half-life	Clinical applications
Technetium (Tc-99m)	140	6 hours	Main SPECT isotope, heart, brain, liver, lungs, cancer imaging
Gallium (Ga-67)	93, 185, 300	3.3 days	Abdominal infections, cancer imaging
Indium (In-123)	171, 245	2.8 days	Infections, cancer imaging
Iodine (I-131)	364	8 days	Thyroid
Thallium (Tl-201)	70, 167	3 days	Myocardial perfusion
Lutetium (Lu-177)	113, 208	6.6 days	Prostate cancer imaging
Actinium (Ac-225)	440	10 days	Emerging tracer, future applications

Dynamic imaging allows the user to investigate temporal changes in activity distribution, such as uptake of the tracer by an organ and its washout (e.g., studies of kidney function, lung ventilation) to study processes related to body physiology. A series of planar or tomographic images are acquired over time.

Gated studies are done by synchronizing photon acquisition from, for example, cardiac perfusion study with an ECG signal and/or with output signal from a monitoring device which tracks patient breathing movements. This allows for creation of 2D or 3D series of images, each corresponding to a particular phase of the moving organ. They can be viewed separately or as a movie.

4.3 Positron Emission Tomography (PET)

In positron emission tomography (PET) imaging, a radioisotope is administered to the patient, which emits β^+ particles (positrons). Emitted positron annihilates with an electron, resulting in simultaneous emission of two 511 keV photons in opposed directions. Radiation detectors are distributed around the patient. If two of them coincidentally detect two photons, it is assumed that the photons originate from the same annihilation event. It is also assumed that the annihilation event has occurred on the line connecting the two detectors (LOR, line of response). This information is a base for image creation.

Scintillators, which are used in PET as detectors, need to have a good time response with a very fast decay of light pulse to allow coincidence detection. They also need to have a good detection efficiency for 511 keV photons, which is usually associated with high density and high atomic number. Materials used in radiography (e.g., CsI, the density of 4.5 g/cm^3) or computed tomography (e.g., CWO, the light decay time of 14.5 μs) are not well suited for that task, compared to, e.g., LSO or LYSO scintillators (density 7–7.5 g/cm^3, decay time ca. 40 ns). With a fast scintillator and a fast light detector (e.g., silicon photomultiplier), it is even possible to estimate time between detection of the two coincidence photons and to approximately determine the position of annihilation along the LOR. Inclusion of the additional data in image reconstruction improves image quality in the so-called time-of-flight (TOF) PET scanners [34].

PET imaging modality for human body applications requires time-of-flight (TOF) correction, which in turn requires precise detection of the time arrival of the photon. Unfortunately, high-Z materials have time uncertainty of the order of nanosecond, while hundreds of picosecond accuracy are required for TOF implementation. For that reason, high-Z materials are currently not used in PET although some research has been going on to resolve that limitation [35].

4.4 Multi-Mode Imaging Modalities

X-ray and gamma ray imaging modalities can be combined together or with other imaging modalities like MRI or ultrasound. Combination of multiple modalities creates even more powerful imaging solution albeit at the higher costs of imaging equipment and related operating cost of administrating the procedure. Examples include combining SPECT with CT to create SPECT functional image superimposed on the CT anatomical image or PET-MRI where functionality of PET is superimposed on MR image. Other combinations, SPECT-MRI, PET-ultrasound, etc., have been also developed. Due to associated cost (the most sophisticated scanner can cost over $4 million a piece), this type of equipment is rarely used in practice except in leading research hospitals in developed countries. More details about them can be found in selected books [9, 13, 15]. Here we are just providing one example of recent SPECT-CT scanner that performs dual-modality function as it is excellent conclusion to the X-ray and gamma ray detection chapter.

One way of overcoming extremely high costs of dual modality equipment is providing anatomical image using low-cost CT systems, e.g., with small number of slices, like 16 or 64 instead of state-of-the-art 320 high-end CT machines while providing functions by nuclear medicine modality. An example of such development is VERITON SPECT-CT scanner that was announced at the EANM 2017 tradeshow in Vienna, Austria, and fully launched in SNMI tradeshow in Philadelphia, USA, in June 2018, as shown in Fig. 5. It is world's first CZT detector-based multi-organ scanner that offers unparalleled sensitivity, image quality, and diagnostic accuracy. Detectors specifically configured for each organ for optimal results.

The purpose of the VERITON-CT scanner is to provide full-body diagnostics driven primary by oncology needs. As a result, the technical requirements for VERITON are somewhat different than for D-SPECT cardiology camera although use CZT detector technology. One obvious difference is requirement in VERITON to have CT image, in order to precisely localize activity spots. Another is the detection area; VERITON-CT requires larger field of view to fit the whole body as illustrated in Fig. 5. The third reason is a range of isotopes used in imaging; while Tc99m is a workhorse contrast used in SPECT, full-body imaging might benefit from other isotopes (possibly used simultaneously with Tc99m). The required energy range of detection is 70 keV (for TI-201 isotope) to 364 keV (for I-131 isotope).

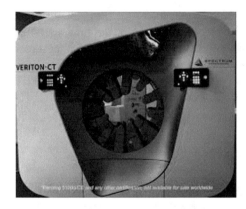 

Fig. 5 SPECT-CT (VERITON-CT) (left) vs SPECT-only (D-SPECT) (right) SPECT cameras from Spectrum Dynamics (https://spectrum-dynamics.com/)

5 Conclusions

X-ray and gamma ray imaging is a well-known imaging modality that has been used for over 100 years since Roentgen discovered X-rays based on his observations of fluorescence. His initial results were published in 1895, and reports of diagnoses of identified fractures shortly followed. A year later, equipment manufacturers started selling X-ray equipment.

Today, X-ray and its three-dimensional (3D) extension, computed tomography (CT), are used commonly in medical diagnosis and are present in all hospitals around the globe [9, 15]. As X-rays are high-energy photons, their generation creates incoherent beams that experience insignificant scatter when passing through various media. As a result, X-ray imaging is typically based on through transmission and analysis of the resulting X-ray absorption data.

Various types of passive and active radiation detectors are used in medical imaging. In high-Z material detectors, energy of X-ray radiation is converted directly to electric charge and then to a digital image. In many older systems, an intermediate conversion of the signal carriers takes place, e.g., to visible light. Examples of the readout detection hardware implementation are illustrated in Figs. 6 and 7.

Different detector materials and designs can be chosen depending on the energy of radiation. The perfect detector could be characterized with high detection efficiency, high spatial, time, and energy resolution. Usually, there is an interplay between the parameters, and depending on the application one may want to, e.g., increase detector's sensitivity at the cost of its spatial resolution. While X-ray imaging in medicine is well-established, the availability of new detectors opens new possibilities. The transition to digital X-ray detection using high-Z materials has only just begun.

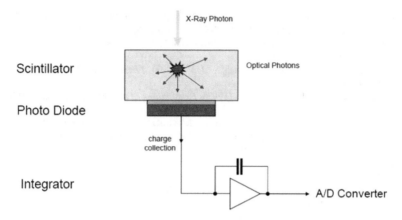

Fig. 6 Radiation detection hardware for indirect detection

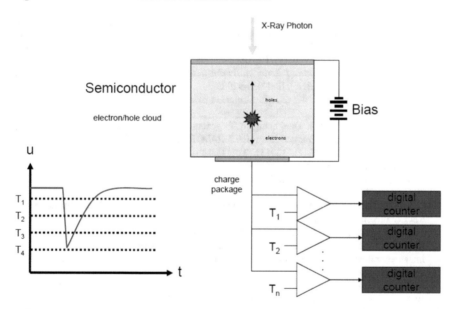

Fig. 7 Radiation detection hardware for direct detection

Bibliography

1. Balter, S. (2019). Fluoroscopic technology from 1895 to 2019 drivers: Physics and physiology. *Medical Physics International, 7*, 111–140.
2. Barber, W. C., Nygard, E., Iwanczyk, J. S., et al. (2009). Characterization of a novel photon counting detector for clinical CT: Count rate, energy resolution, and noise performance. *Proceedings of SPIE, 7258*, 725824.
3. Bazalova, M., Kuang, Y., Pratx, G., & Xing, L. (2012). Investigation of X-ray fluorescence computed tomography (XFCT) and K-edge imaging. *IEEE Transactions on Medical Imaging, 31*, 1620–1627. https://doi.org/10.1109/TMI.2012.2201165

4. Bazalova-Carter, M., Ahmad, M., Matsuura, T., et al. (2015). Proton-induced x-ray fluorescence CT imaging. *Medical Physics, 42*, 900–907. https://doi.org/10.1118/1.4906169
5. Behling, R. (2018). X-ray tubes development – IOMP history of medical physics. *Medical Physics International, 6*, 8–55.
6. Cherry, S. R., Sorenson, J. A., & Phelps, M. E. (2003). *Physics in nuclear medicine*. W. B. Saunders, Elsevier Science.
7. Cho, Z.-H., Jones, J. P., & Singh, M. (1993). *Foundations of medical imaging*. John Wiley & Sons.
8. Cowen, A. R., Davies, A. G., & Kengyelics, S. M. (2007). Advances in computed radiography systems and their physical imaging characteristics. *Clinical Radiology, 62*, 1132–1141. https://doi.org/10.1016/j.crad.2007.07.009
9. Farncombe, T., & Iniewski, K. (2013). *Medical imaging technology and applications*. CRC Press. ISBN 9781466582620.
10. Fuchs, T., Kachelrie, M., & Kalender, W. A. (2000). Direct comparison of a xenon and a solid-state CT detector system: Measurements under working conditions. *IEEE Transactions on Medical Imaging, 19*, 941–948. https://doi.org/10.1109/42.887841
11. Grajo, J. R., Patino, M., Prochowski, A., & Sahani, D. V. (2016). Dual energy CT in practice. *Applied Radiology, 45*, 61–62.
12. Green, F. H., Veale, M. C., & Wilson, M. D. (2016). Scatter free imaging for the improvement of breast cancer detection in mammography. *Physics in Medicine and Biology, 61*, 7246–7262.
13. Iniewski, K. (2009). *Medical imaging: Principles, detectors, and electronics*. Wiley. ISBN: 978-0-470-39164-8, https://doi.org/10.1088/0031-9155/61/20/7246
14. Iniewski, K. (2022). *Advanced X-ray radiation detection: Medical imaging and industrial applications*. Springer. ISBN: 978-3-030-92989-3.
15. Iwanczyk, J., & Iniewski, K. (2015). *Radiation detectors for medical imaging*. CRC Press. ISBN 9781498704359.
16. Johnson, T. R. C. (2012). Dual-energy CT: general principles. *AJR. American Journal of Roentgenology, 199*, 3–8. https://doi.org/10.2214/AJR.12.9116
17. Jones, B. L., & Cho, S. H. (2011). The feasibility of polychromatic cone-beam x-ray fluorescence computed tomography (XFCT) imaging of gold nanoparticle-loaded objects: A Monte Carlo study. *Physics in Medicine and Biology, 56*, 3719–3730. https://doi.org/10.1088/0031-9155/56/12/017
18. Karçaaltıncaba, M., & Aktaş, A. (2011). Dual-energy CT revisited with multidetector CT: Review of principles and clinical applications. *Diagnostic and Interventional Radiology, 17*, 181–194. https://doi.org/10.4261/1305-3825.DIR.3860-10.0
19. Kaufman, L., & Carlson, J. (2011). An evaluation of airport x-ray backscatter units based on image characteristics. *Journal of Transportation Security, 4*, 73–94. https://doi.org/10.1007/s12198-010-0059-7
20. Kemerink, G. J., Kütterer, G., Kicken, P. J., et al. (2019). The skin dose of pelvic radiographs since 1896. *Insights Into Imaging, 10*, 39. https://doi.org/10.1186/s13244-019-0710-1
21. Kowalski, P., Wiślicki, W., Shopa, R. Y., et al. (2018). Estimating the NEMA characteristics of the J-PET tomograph using the GATE package. *Physics in Medicine and Biology, 63*, 165008. https://doi.org/10.1088/1361-6560/aad29b
22. Lai, C. J., Shaw, C. C., Geiser, W., et al. (2008). Comparison of slot scanning digital mammography system with full-field digital mammography system. *Medical Physics, 35*, 2339–2346. https://doi.org/10.1118/1.2919768
23. Lell, M. M., Wildberger, J. E., Alkadhi, H., et al. (2015). Evolution in computed tomography: The battle for speed and dose. *Investigative Radiology, 50*, 629–644. https://doi.org/10.1097/RLI.0000000000000172
24. Machida, H., Yuhara, T., Tamura, M., et al. (2016). Whole-body clinical applications of digital tomosynthesis. *Radiographics, 36*, 735–750. https://doi.org/10.1148/rg.2016150184
25. Moskal, P., Rundel, O., Alfs, D., et al. (2016). Time resolution of the plastic scintillator strips with matrix photomultiplier readout for J-PET tomograph. *Physics in Medicine and Biology, 61*, 2025–2047. https://doi.org/10.1088/0031-9155/61/5/2025

26. Patel, B. K., Lobbes, M. B. I., & Lewin, J. (2018). Contrast enhanced spectral mammography: A review. *Seminars in Ultrasound, CT, and MR, 39*, 70–79. https://doi.org/10.1053/j.sult.2017.08.005
27. Redler, G., Jones, K. C., Templeton, A., et al. (2018). Compton scatter imaging: A promising modality for image guidance in lung stereotactic body radiation therapy. *Medical Physics, 45*, 1233–1240. https://doi.org/10.1002/mp.12755
28. Samei, E., & Flynn, M. J. (2003). An experimental comparison of detector performance for direct and indirect digital radiography systems. *Medical Physics, 30*, 608–622. https://doi.org/10.1118/1.1561285
29. Shefer, E., Altman, A., Behling, R., et al. (2013). State of the art of CT detectors and sources: A literature review. *Current Radiology Reports, 1*, 76–91. https://doi.org/10.1007/s40134-012-0006-4
30. Sprawls, P. (2018). Film-screen radiography receptor development – A historic perspective. *Medical Physics International, 6*, 56–81.
31. Stepusin, E. J., Maynard, M. R., O'Reilly, S. E., et al. (2017). Organ doses to airline passengers screened by X-ray backscatter imaging systems. *Radiation Research, 187*, 229–240. https://doi.org/10.1667/RR4516.1
32. Taguchi, K., & Iwanczyk, J. S. (2013). Vision 20/20: Single photon counting x-ray detectors in medical imaging. *Medical Physics, 40*, 100901. https://doi.org/10.1118/1.4820371
33. Udoye, C. I., & Jafarzadeh, H. (2010). Xeroradiography: Stagnated after a promising beginning? A historical review. *European Journal of Dentistry, 4*, 95–99. https://doi.org/10.1055/s-0039-1697816
34. Vandenberghe, S., Mikhaylova, E., D'Hoe, E., et al. (2016). Recent developments in time-of-flight PET. *EJNMMI Physics, 3*, 3. https://doi.org/10.1186/s40658-016-0138-3
35. Wang, Y., & Abbaszadeh, S. (2022). Optical properties modulation: A new direction for the fast detection of ionizing radiation in PET. In K. Iniewski (Ed.), *Advanced X-ray radiation detection: Medical imaging and industrial applications*. Springer.
36. Webb, S. (Ed.). (1988). *The physics of medical imaging*. Institute of Physics Publishing.
37. Yan, H., Tian, Z., Shao, Y., et al. (2016). A new scheme for real-time high-contrast imaging in lung cancer radiotherapy: A proof-of-concept study. *Physics in Medicine and Biology, 61*, 2372–2388. https://doi.org/10.1088/0031-9155/61/6/2372

Large Area Thin-Film CdTe as the Next-Generation X-Ray Detector for Medical Imaging Applications

Fatemeh Akbari, E. Ishmael Parsai, and Diana Shvydka

1 Introduction

Medical imaging relies on the detection of x-ray and γ-ray radiation. Planar radiography and mammography, computed tomography (CT), as well as multiple nuclear medicine applications such as planar gamma cameras, single-photon emission computed tomography (SPECT), and positron emission tomography (PET) are prominent examples. Most of these imaging systems employ an indirect detection configuration, where a high-energy source particle interacts with a scintillator or phosphor layer, generating optical photons which in turn are collected by either photomultiplier tubes or, more recently, semiconductor photodetectors. Alternatively, semiconductor devices could be used in a direct detection design as their sensitive volume may serve both purposes: to generate electron-hole pairs upon a high-energy photon absorption and collect the charge carriers at the device terminals with application of a built-in or applied electric field. In this context, semiconductor detectors have a key role in all modern digital imaging applications. The success of semiconductor materials in radiation detection applications can be attributed to a number of distinctive features not found in other types of devices, including high absorption, low effective ionization energy, availability of pulse mode operation leading to spectroscopy with superior energy resolution, and the ability to construct compact and robust direct detection systems. Even though energy resolution may not be immediately relevant in planar radiography, it is important in most of nuclear imaging and other more advanced (e.g., energy-selective) applications.

Silicon (Si) and germanium (Ge) are the two most widely utilized materials, employed in charged particle and γ-ray detectors with high-resolution spectroscopy

F. Akbari · E. I. Parsai · D. Shvydka (✉)
Department of Radiation Oncology, University of Toledo HSC, Toledo, OH, USA
e-mail: Diana.Shvydka@utoledo.edu

© The Author(s), under exclusive license to Springer Nature Switzerland AG 2023
L. Abbene, K. (Kris) Iniewski (eds.), *High-Z Materials for X-ray Detection*,
https://doi.org/10.1007/978-3-031-20955-0_2

capabilities. Due to their narrow band gap, however, both require cooling to achieve satisfactory signal-to-noise ratio. Additionally, their rather low absorption efficiency created an ongoing demand to design compact room temperature detectors with significantly improved efficiency and energy resolution [1]. Semiconductor detector technology has recently progressed to the point that it can now replace scintillators [2], especially with development of compound semiconductors, greatly expanding material choices. The focus had also switched from growth of small area crystals to deposition of continuous polycrystalline films, making large area detectors economically feasible. The latter development was prompted by rapid progress in thin-film solar cell technology utilizing high-Z compound semiconductors, having properties suitable for high-performance radiation detectors.

Solar cells operating principle is based on the photovoltaic (PV) effect, involving creation of an electron-hole (e-h) pair upon absorption of incident photons having energy above the semiconductor band gap. In radiation detection applications, these free carriers are generated in response to high-energy x-rays, producing thousands of e-h pairs (in proportion to much larger energy transferred to the medium) instead of a single one. Most radiation detectors operate under reverse bias to enhance free carrier separation, but can also be utilized in photovoltaic mode with no bias. The most important difference is their thickness: typical semiconductor radiation detectors have ~1 mm thickness, while solar cells are in the range of ~1 μm, dictated by the visible light absorption length. Even at that small thickness, photovoltaic devices have been shown to work for high-energy radiation detection, suitable for high dose and high-dose rate radiation sources.

Cadmium telluride (CdTe) is a prominent example of a thin-film solar cell material, developed to yield low-cost, highly efficient photovoltaic devices [3–5], that is very appealing for construction of radiation detectors. Single crystal devices have received great attention from the scientific community involved in x-ray and γ-ray applications in medicine due to their physical properties [6–10]. CdTe has a wide bandgap (1.5 eV), low ionization energy (~4.4 eV per e-h pair), a high bulk resistivity (10^9 Ω) [11], a high atomic number (Cd:48 and Te:52), and a high density (5.85 g/cm^3), in addition to appropriate mobility-lifetime product, dependent on the material quality. CdTe has also been shown to have excellent radiation hardness, making it a perfect choice for imaging applications. All these characteristics make it suitable for room-temperature operation with properties far exceeding those of the classical Si and Ge [12–14]. Figure 1 illustrates a comparison of the mass attenuation coefficient of CdTe, Si, and Ge. Two additional semiconductor materials are included in the figure: selenium (Se), utilized in presently commercially available detector technology, and mercuric iodide (HgI$_2$), representing a compound with attractive absorption properties, but thus far inferior in its achievable electrical properties.

This chapter evaluates the performance of CdTe semiconductor material for use in the next-generation large-area thin-film medical imagers, in both diagnostic and therapeutic energy ranges. We focus on parameters characterizing signal and noise propagation, obtained with Monte Carlo (MC) modeling, discuss device operation, and present some verification measurement results.

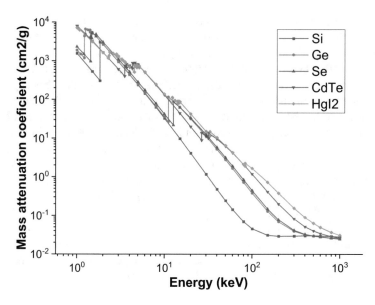

Fig. 1 Mass attenuation coefficients for Si, Ge, Se, CdTe, and HgI_2. Reproduced from NIST data [15]

2 Large Area Thin-Film CdTe Detector for Medical Imaging Applications

2.1 Direct Detection Configuration and Imaging Characteristics

The emergence of digital radiography technology and the following use of flat panel detectors transformed the medical imaging field. The next step of switching to direct detection systems, which would benefit a simpler device structure and lower manufacturing costs, is still slow to be widely implemented. Thus far amorphous selenium (a-Se) is the only photoconductor developed into commercial direct detection medical imagers for both general radiography and mammography [16–19]. However, its low x-ray absorption and high effective ionization energy (~50 eV) result in inadequate sensitivity, especially important for low exposure levels of fluoroscopic mode [20].

Materials with high atomic numbers and densities have been proposed to replace a-Se, such as mercuric iodide (HgI_2), lead iodide (PbI_2), lead oxide (PbO) [21], thallium bromide (TlBr) [22], and cadmium telluride/cadmium zinc telluride (CdTe/CdZnTe) [23–26]. These materials have lower effective ionization energy than a-Se; significantly larger band gaps, required to reduce leakage currents at room temperature; and a high mobility-lifetime product, which allows for effective charge collection [27, 28]. Among them, CdTe appears to be one of the most promising materials. Despite the fact that CdTe has a lower average atomic number than HgI_2, their absorption properties are pretty similar over a large range of keV x-ray energy [15] and only slightly lower in the MeV range.

A quantitative evaluation of a detector is performed through calculation of its detective quantum efficiency (DQE), characterizing the overall efficiency of the detector system in transferring squared signal to noise ratio (SNR) of the incident x-ray beam to the detector output. It accounts for the stochastic nature of x-ray absorption in the material and corresponding fluctuation and noise in the image formed in the detector. Instead of the physical distance x, the signal spreading is expressed through the spatial frequency, often measured in units of cycles per mm (1/mm), defined as $f = 1/(2\Delta x)$, with the low frequency representing the ability of the detector to reproduce large-scale features, while the high frequency is reflective of the small-scale features. The frequency-dependent DQE(f) can be calculated based on an average deposited energy A; the number of incident photons per unit area q_0; modulation transfer function MTF(f); a Fourier transform of line spread function LSF(x), defined in the real space; and noise power spectrum of the energy absorbed within a thin-film CdTe layer NPS(f) as [29]:

$$\text{DQE}(f) = \frac{A^2 * \text{MTF}^2(f)}{q_0 * \text{NPS}(f)} \tag{1}$$

We assumed polycrystalline CdTe layer to be spatially uniform, since the typical grain size of ~1 μm is much smaller than the film thickness, allowing for averaging of all structural nonuniformities. All parameters in Eq. (1) were calculated based on the MC modeling on the energy deposition process with MCNP5 package [30]. The CdTe layer was divided into 512 strips on each side of the slit source, parallel to the source; energy deposition in each strip was collected with *F8 pulse height tally and represented discrete LSF(x). Frequency-dependent MTF(f) was found applying fast Fourier transform to LSF(x) using Hanning window method.

2.2 Diagnostic X-Ray Imaging

In this section, results of our investigation on employing CdTe photoconductor for large-area direct detection system in x-ray diagnostic imaging applications are presented. Based on the measured output of our diagnostic x-ray simulator, we first evaluated the input x-ray spectra for a set of tube potentials ranging from 70 to 140 kVp using the tungsten anode spectral model [31].

Using these spectra as input, MC simulations of energy deposition were conducted for the 10-μm-wide slit, which according to the Nyquist criterion gave a cutoff frequency of 50 mm^{-1}. The MC simulation geometry included a 20-cm-thick water phantom, located at 100 cm source to surface distance, and an air gap with 20 cm width between the phantom and detector surface. The amplification (conversion) gain, associated with conversion of the energy deposited by x-rays into electron-hole pairs in CdTe, was also calculated.

2.2.1   Detector Optimization

Pre-sampling Modulation Transfer Function (MTF)

MTF characterizes the spatial frequency response of an imaging system or a component, representing its ability to transfer signal without spreading and thus defining its resolution [32]. Shown in Fig. 2a, MTF(f) decreased with increasing CdTe thickness, becoming much less pronounced for films thicker than 300 μm. An increase in both secondary particle scatter and the fraction of k-fluorescence x-rays re-absorption is responsible for this general trend. The MTF degrades more at higher frequencies because the probability of Compton interactions increases, resulting in scattered particles depositing their energy close to the first interaction site. The effective path lengths of recoil electrons increase with the increasing energy, resulting in increased lateral spread within the detector. These observations are consistent with the previously published findings [28, 33, 34].

We also compared characteristics of CdTe with a-Se and HgI_2. In a-Se, the lowest absorption, which leads to the lowest interaction probability, results in the lowest signal spreading and higher MTF. Despite having the highest atomic number and hence the highest absorption, HgI_2 has a greater MTF for the 80 kVp x-ray spectrum than CdTe, which could be explained by the longer path length of fluorescence photons produced from x-rays with energy slightly above the k-edges of both Cd and Te [28]. At 120 kVp, however, where the incoming beam contains more photons with energies above the HgI_2 k-edge (83 keV), this is no longer the case: CdTe has a greater MTF than mercuric iodide.

Noise Power Spectra (NPS)

Quantum noise is an unavoidable result of statistical fluctuations in the number of x-rays interacting within the detector and statistical fluctuations in the number of electrons produced as the result of each interaction. NPS describes both the magnitude and spatial frequency characteristics of image noise [35, 36].

The noise power increased with increasing energy for the same film thickness, most likely due to an increase in energy deposited per interacting photon. Another observed trend was strong correlation of the noise. Secondary particles, produced by interacting x-rays, deposit energy in the CdTe layer through a large set of discrete interactions along each particle's trajectory, resulting in correlated quantum noise. Additionally, when the detector thickness increased, more photons were absorbed, and the NPS increased as well, as shown in Fig. 2b. For 300 to 1000-μm-thick CdTe, typical values were close to 0.1 mm, which is within the range of typical pixel sizes used in digital imagers [20]. Thus the noise correlation would be contained withing a pixel.

Detective Quantum Efficiency (DQE)

DQE has become the best single descriptor of an imaging detector performance. It is equal to 1 in an ideal imaging system, degraded in reality by various sources of noise correlated with the system [37]. As shown in Fig. 2c, the DQE(f) improves

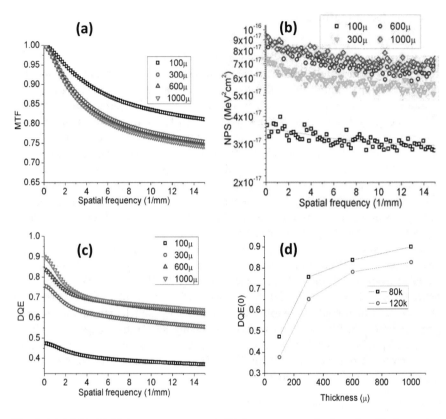

Fig. 2 (**a**) MTF; (**b**) Noise power spectra (solid lines represent the best fit with Lorentzian function); (**c**) DQE calculated with fitted NPS under 80 keV. CdTe thicknesses are 100, 300, 600, and 1000 μm and (**d**) DQE(0) of CdTe detector vs. the film thickness in μm for 80 and 120 keV x-ray spectra

with increasing thin-film CdTe thickness due to increase in number of absorbed photons. A thicker detector, on the other hand, gives longer paths for secondary electrons and photons to spread laterally, resulting in a loss of spatial resolution and an increase in noise.

DQE(0), detective quantum efficiency due to energy absorption, describes the ability of an imaging system to transfer information content of an input signal. It is commonly used as a quantitative reference for evaluating a detector system. It has been demonstrated [38, 39] that DQE(0) can be defined in terms of moments of spectral absorbed energy distribution (AED), which makes MC simulations of different detector designs comparatively quick. We presented DQE(0) values derived from DQE(f) analysis in Fig. 2d to describe the patterns in energy absorption with thickness and the source energy, defined by kVp. We can readily see how the detector's absorption efficiency increases as the thickness increases, effectively saturating after 600 μm, especially for lower energies.

Conversion Gain (g)

Detector gain is associated with conversion of the energy deposited by x-rays into electron-hole pairs in CdTe. The amount of energy deposited in CdTe per incoming x-ray is defined by the first moment of AED and therefore depends on CdTe thickness, as well as the converter layer material and thickness. The maximum amplification gain g can be estimated as $g = $ AED/W_{CdTe} and represents the upper limit estimate for the number of electron-hole pairs generated in a certain volume of CdTe. To evaluate the conversion gain, we estimated the number of electron-hole pairs created in the CdTe layer per 1 cm^2 area of the detector using the energy required for creating one electron-hole pair in CdTe for polycrystalline material [40, 41]. We found dependences of generation rates on the CdTe thickness close to $(1 - \exp(-\alpha t))$ up to the CdTe thickness t of the order of the average x-ray penetration depth, saturating for thicker films.

2.2.2 Mammography Application

CdTe has been studied as an optimum semiconductor sensor material for mammography compared to other materials. To achieve acceptable detection efficiency for mammography, CdTe of 450 μm is sufficient, but much thicker Si is required for its less attenuation efficiency, which in turn leads to a lower contrast and image sharpness. Although GaAs has a higher mass attenuation efficiency in the mammography energy range, it has a poor image sharpness and lower contrast to noise ratio (CNR) and spatial resolution due to x-ray fluorescence generation with high yield resulting from its low k-edge. Moreover, an experimental measurement on a human breast tissue phantom for comparison between the clinical a-Se detector and a photon processing semiconductor-based detector with CdTe- and Si-sensors showed a higher CNR is achievable using CdTe [42].

2.3 Radiation Therapy Imaging

Verification of the patient's positioning before radiation therapy is required for the accurate delivery of prescribed tumor radiation dose in external beam radiotherapy. In modern linear accelerators (linacs), equipped with keV imaging source and detector, it is usually accomplished with acquisition of a cone beam CT scan (CBCT), which is compared to a diagnostic CT scan used for treatment planning. In older linacs, it is done utilizing electronic portal imaging devices (EPIDs), which still come standard due to their extensive use in linac and patient-specific quality assurance (QA) and dosimetry applications. The size of the EPID is dictated by the size of the largest treatment field, which is usually 40×40 cm^2 at the isocenter. Hydrogenated amorphous silicon (a-Si:H) is used in most common commercial EPID systems in an indirect detection design, where high-energy x-rays are absorbed in a scintillating layer, generating optical photons which are then detected by a-Si:H photodiodes [43]. Given the treatment field dimensions, use of polycrystalline or amorphous semiconductors is preferable to their single-crystal counterparts. Silicon is unsuitable for direct x-ray conversion-type detectors due to low atomic number and electron density, resulting in low quantum efficiency and poor radiation hardness [44]. Although significant attempts have been made to increase x-ray absorption by using a very thick (>10 mm) detection medium [45–47], the resulting devices have a very complex design and can be extremely expensive to manufacture.

A very cost-effective alternative approach is using high electron density semiconductor thin films such as a-Se [48–50] and metal halide compound semiconductors [51, 52] in direct detection design. While a-Se has already been utilized in commercially available diagnostic energy range detectors, the quantum efficiency for MeV range limited the widespread use of this material. Despite their high electron densities and average atomic number, thin-film metal halides, such as HgI_2 and PbI_2, have not yet been implemented into commercial devices, mainly due to poor material quality: their grain boundaries serve as efficient recombination centers for the generated electron-hole pairs. In contrast to this general behavior, thermal annealing in the presence of $CdCl_2$ produces benign grain boundary features in polycrystalline CdTe thin films, developed for photovoltaic applications [53]. The advantage of the small detector thickness is that the low hole mobility and the electron trapping, which are known limiting factors in crystalline CdTe detectors, become much less important as drift time gets smaller than the recombination time.

2.3.1 Detector Design and Optimization

In radiation therapy, a linac photon beam is characterized by the nominal energy of electrons striking the accelerator target, for example, 6 MeV. Bremsstrahlung interactions of such electrons with high atomic number material of the target result in the production of polyenergetic x-ray beam, having the average energy of 2 MeV, and delivered through pulses with typical repetition frequency of 400 Hz and a dose rate of 600 cGy/min. As with the diagnostic imaging energy range, discussed in the previous section, Monte Carlo simulation [54] was used to study the design consideration of a large area thin-film CdTe-based imaging system. At this photon energy, the absorption by a thin CdTe layer is fairly low, but can be augmented by adding a metal layer of high atomic number and density functioning as a converter.

Benefit of such a converter is the absorption of the low-energy noise in the x-ray spectrum generated by the linac and scattering from the patient [55]. The proposed design utilizes direct detection, which allows for noise reduction and better image resolution [43], as opposed to the commercially available indirect type EPIDs [56].

Using MCNP5 package and SCAPS-1D software [57], we evaluated the performance of the detector by first modeling the head of a linac, providing the realistic photon treatment source, 20 cm water phantom representing a patient, and thin-film CdTe combined with a metal plate. The photon beam spectrum was calculated at a typical source-to-detector distance of 172 cm, with the water phantom placed at

100 cm source to surface distance (thus leaving a 52-cm-wide air gap between the phantom and the detector surface). Since copper (Cu) metal plate is typically used with a commercial EPID system, we considered Cu plate in combination with CdTe layers. For simulation details, see [58].

DQE

We found that DQE(0) values calculated for a set of 300-μm-thick sensor materials combined with a Cu metal plate of varying thickness under 6 MeV beam show that CdTe offers about 18 relative percent increase compared to a-Se and phosphor screen typically used in direct and indirect detection configuration, respectively. As illustrated in Fig. 3b, as the Cu thickness increases, the corresponding DQE(0) for each material raises dramatically, achieving a saturation for copper thickness of ~3 mm, which corresponds to the average secondary electron range in Cu. This optimal thickness, corresponding to the highest value of DQE(0), was determined for other converter plate metals (Al, Cu, Pb, and W). Aluminum was found to offer the highest enhancement in DQE(0). However, its large physical thickness contributes to significant signal spreading. These results are in good agreement with those previously reported for different absorbers [49, 59].

MTF

We evaluated LSF(x) and corresponding MTF(f) in the narrow slit geometry for CdTe layers of different thickness in combination with 3 mm copper converter plate. MTF curves in Fig. 3c indicate that the thicker films introduce greater spreading due to the increased lateral path of secondary electrons, having an average range of ~1–2 mm in CdTe at MeV energies [15].

NPS

Simulation results showed that noise is strongly correlated, especially for realistic detector configurations with metal converter plates. The correlation length in mm reflects the smallest feasible pixel size for the detector, but is still below the lower boundary of typically accepted pixel size for megavoltage imagers. For example, 1-mm-thick semiconductor layer with 3 mm Cu plate configuration results in correlation length of 0.37 mm (cf. typical EPID pixel size of 0.625 mm).

Device Operation

We evaluated the detector operation by modeling its current-voltage ($I–V$) characteristics using SCAPS-1D software package originally developed for thin-film solar cells [57]. The sketch of a modeled device is shown in Fig. 4a. The electron-hole generation profiles in CdTe films of various thicknesses were created by dividing MC-simulated profiles of energy deposited across the device thickness by the average ionization energy in CdTe ($W_{CdTe} = 4.43$ eV for crystalline and 5 eV for polycrystalline material). These profiles were used as an input for SCAPS to model the expected device open-circuit voltages (V_{OC}) [57].

It was observed that the output voltage does not change dramatically with increasing CdTe thickness, even though the energy absorbed in the device increases.

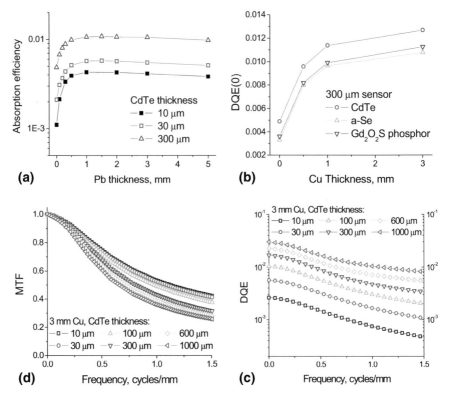

Fig. 3 (**a**) Absorption efficiency of 10-, 30-, and 300-μm-thick CdTe layer obtained by MC simulations for 6 MeV photon spectrum with addition of metal (lead) converter of varying thickness [55]; (**b**) DQE(0) for three sensor materials, 300 μm thick, in combination with Cu plate of varying thickness; (**c**) Modulation transfer function for CdTe of different thickness with 3-mm-thick Cu metal plate; (**d**) Comparison of DQE(*f*) for CdTe of different thickness with 3-mm-thick Cu metal plate

Another important parameter used in the device modeling, which directly affected charge carrier recombination and charge trapping in the material, was the density of defects in CdTe layer. Device performance degraded with increasing defect density. V_{OC} was calculated in the range of tenths of a volt depending on the material quality through the defect concentration N_d, but almost independent of thickness. The estimated current density was approximately six orders of magnitude lower than the typical 1 sun (100 mW/cm^2) illumination short circuit current of ~20 mA/cm^2 [5]. This means that the x-ray energy absorbed by the device is in the micro-sun range. Therefore, current measurements appear to be less trustworthy, making voltage reading more advantageous approach. The current change was linear with energy deposition (and CdTe thickness), while the voltage increase was only logarithmic, following a typical exponential diode *I–V* curve dependance, shown in Fig. 4b.

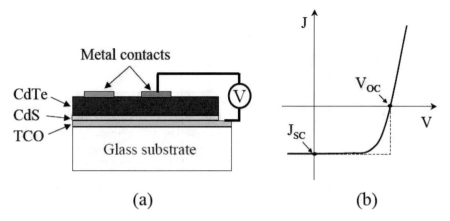

Fig. 4 (a) Sketch of a solar cell (not to scale) and (b) typical current-voltage curve with major cell parameters identified

The signal to noise ratio, however, was greatly improved for thicker CdTe layers as demonstrated by increase in DQE with thickness, as shown in Fig. 3d.

2.3.2 Verification Measurements

Our calculations were also verified using CdS/CdTe solar cells with 10-μm-thick CdTe layer [55]. They were fabricated by the sequential deposition of SnO_2, CdS, CdTe, and a metal back contact layer on 1.5 × 1.5 in glass substrate [56]. The device structure shown in Fig. 4a includes a very thin CdS layer resulting in built-in electric field ~10^4 V/cm and thus was used in photovoltaic mode (no external biasing). All of the cells demonstrated efficiencies in 12% range and V_{OC} values above 800 mV under standard 1 sun illumination. Because the devices were sensitive to light, they were wrapped in black cloth before each measurement.

The temporal response of the CdTe detector was verified under the 6 MeV pulsed x-ray beam demonstrating the output signal adequately following the change in the irradiation intensity, as depicted in Fig. 5a. The associated RC time parameter increased with the dose rate, consistent with the inherent increase in the device resistance R induced by the built-in charge.

Simple output voltage acquisitions with a digital multimeter and oscilloscope served to determine the depth dose distribution in a polystyrene phantom under 6 MeV linac beam; the results compared very favorably to both MC simulations and measurements with the clinical 0.6 cm^3 Farmer ion chamber (the standard dosimetry tool), as shown in Fig. 5b.

The measured output signal values in the range of ~0.1 V were lower than predicted by SCAPS modeling (close to 0.2 V) under 6 MeV x-ray beam of medical linac. For comparison, a silicon detector operating in photovoltaic mode under Co-60 photon source produced an open-circuit voltage in the range of 0.01 V [60, 61].

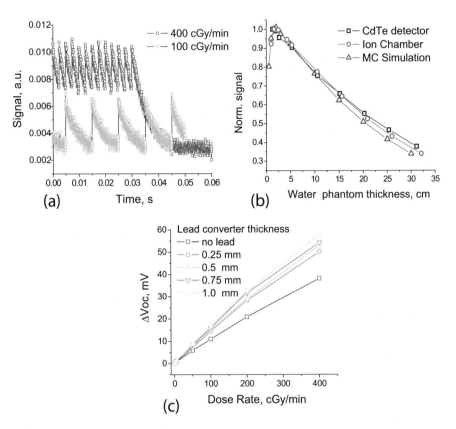

Fig. 5 (**a**) Time-dependent response of CdTe detector for two dose rates; (**b**) Normalized output signal versus polystyrene phantom thickness measured with CdTe detector and ion chamber, compared with results of MC simulations; (**c**) Increase in open-circuit voltage versus dose rate. "no-lead" trend in the figure represents measurement with the original thick back contact, already acting as thin metal converter [55]

Experiments verifying the advantage of the metal converter indeed showed it to significantly boost the measured V_{OC}, especially for higher linac dose rates. As evident from Fig. 5c, V_{OC} is practically linear in signal intensity (dose rate) due to very low excitation rate far below the standard 1 sun condition where the standard logarithmic expression [62] for V_{OC} can be linearized. The signal was corrected for the background, setting the dark (0 dose rate) point to 0.

Overall, we conclude that the proposed design utilizing a photovoltaic device, with a built-in electric field [63], paves the way to a promising large area, inexpensive x-ray detector possessing superior radiation hardness, discussed in Sect. 3.

2.3.3   Multilayer Detector

A further enhancement of the design with a top metal converter plate had been explored through investigation of electron backscattering phenomenon, known to be highly dependent on the material atomic number Z. Adding an electron reflector in tandem with the back electrode was found to enhance the energy deposition and the corresponding detector signal in CdTe layer by ~75% for lead (Pb) under high-energy x-rays. Inserting a very thin low Z material between the top metal plate and the semiconductor layer decreased secondary electron backscattering leading to more of them entering the sensitive layer and also contributing to signal enhancement [64].

An interesting new development in field of high-energy radiation detectors attempts to overcome the problem of low absorption by thin detectors and signal spreading with the thick ones by building a stacked multilayered device. This configuration allows for independent signal reading of each layer within a stack, which tremendously increases the amount of collected information. They are capable of high-energy photon imaging and spectroscopy and subsequent extensions to Compton camera operations and arbitrary particle detection/spectroscopy. For example, recently developed Si/CdTe Compton cameras that use one-layer Si of 500 μm and a stack of three CdTe layers of 750 μm thickness have already been tested in medical imaging. Some applications include double-photon emission imaging [65], imaging ^{99m}Tc-DMSA accumulated in rat kidneys [66], and simultaneously imaging the distributions of ^{99m}Tc-DMSA and 18F-FDG in a human body [67]. All these examples relied on crystalline semiconductors. However, the possibility of utilizing thin-film CdTe devices with 15 to 30 layers and independent signal readouts presents an opportunity to enhance the image quality with regularization. Thus far this approach had been investigated only theoretically [68].

3   Radiation Effects on CdTe Detector

The effects of radiation on CdTe thin films have been studied for various radiation sources. Available literature indicates structural damage to the film under protons, heavy ions, and neutrons due to atomic displacements [69–74]. High-energy photon sources, however, are not expected to cause the same changes withing the detector material. We can evaluate the probability of atomic displacements in CdTe film under a 6 MeV linac photon beam with a simple estimate. The primary photon interaction at this energy range is Compton scattering, resulting in the energy transfer to an electron, set in motion after the interaction. The energy transferred to an atom is smaller than the electron kinetic energy by the ratio of the electron to atom masses, $\sim 4 \times 10^{-6}$. For the x-ray source (the effective energy of 2 MeV and maximum energy 6 MeV), the average electron energy received by Compton electron is approximately 50% of the source photon energy [75] and is never higher than its maximum energy. As a result, the atom receives <4 eV on average and cannot receive more than 24 eV. Comparing these values with those of the minimum displacement threshold energy estimated for Cd and Te as 19 and 15 eV, respectively [76], makes atomic displacements possible in principle, while their probability is vanishingly small. Other mechanisms of energy transfer are even less efficient. This estimate illustrates that for diagnostic photon sources, having energy ~100 keV, still dominated by Compton interactions, the atomic displacements become unrealistic.

The first experimental investigation of radiation hardness under x-ray beam of therapeutic energy range studied the stability of thin-film CdTe devices utilizing 6 MeV photon beam of Elekta Precise SL-25 linear accelerator [77]. The study utilized solar cell devices with 3-μm-thick CdTe fabricated on 3 mm glass substrate, sketched in Fig. 4a [78, 79]. In this work, CdTe samples received doses of up to 25 kGy, exceeding the typical dose received by EPID system during its lifetime. Cell efficiency and other PV parameters were monitored after each irradiation session and were found to decrease mainly due to darkening of the glass substrate. After glass transmission loss correction, the efficiency dependence on dose absorbed in the CdTe layer demonstrated no consistent trend of efficiency degradation or change in sample non-uniformity. A decrease in transmittance of the glass substrate was also reported by other researchers when a Co-60 gamma-ray source was employed and the total gamma irradiation dose was more than 500 kGy.

Experiments with cells deposited by different techniques, such as sputtering and vapor transport deposition, led to the same conclusion of superior thin-film CdTe stability under high-energy x-ray irradiation [80]. More recent studies extended the range of the irradiation doses even further. The optical properties of ultra-thin, 50 nm, CdTe film were found to decrease with the increase of the dose of γ-irradiation (Co-60 source) up to 120 kGy [81]. Another study found that the 5-μm-thick CdTe solar cells had sufficient tolerance against Co-60 γ-ray exposure, taking the devices to a much higher total dose of up to 3 MGy [82]. Several groups had noted some evidence of structural changes in the investigated devices, such as metal contact cracking or delamination, making it harder to distinguish between material and device component degradation.

The remarkable CdTe stability under high-energy x-ray irradiation could be explained by relative inefficiency of two leading mechanisms of semiconductor material degradation under irradiation. The first is the lattice defect creation through atomic displacements, discussed above, at its worst leading to structural defects that in turn can promote metal diffusion from contacts over very long exposure times. The second degradation mechanism, well established in PV devices, is more specific to polycrystalline and amorphous materials and occurs through electron-hole pair generation [83] and their subsequent effect on the localized electronic states [41]. Taking into account very low electron-hole pair generation rate under typical radiation treatment accelerator beam (several orders of magnitude lower than under 1 sun light intensity), the second mechanism is also not expected to contribute to device degradation until extremely high radiation levels are reached. While the studies were conducted under MeV photon energy beam, similar results could be expected for the keV photon sources employed in the diagnostic imaging.

4   Conclusions

To conclude, in all diagnostic imaging applications the CdTe-based detector is capable of achieving a high-resolution as well as high-quantum efficiency. The optimal thickness of thin-film CdTe for diagnostic x-ray imaging should be in the range of 300–600 μm.

Utilization of high-density and high-atomic number CdTe detector is especially important for lowering the exposure levels of fluoroscopic imaging mode, since they are capable of not only detecting a larger fraction of the incident radiation, but are also likely to reduce the lateral spread of secondary electrons and photons, resulting in higher-quality images in contrast and resolution.

In performing DQE(f) and scatter analysis in design of the proposed CdTe detector for MeV energy range applications, the detector of realistic semiconductor thickness below 3.7 mm should always be used with the metal converter to enhance x-ray absorption.

We have enhanced the traditional detector modeling with a new step translating the Monte Carlo simulated energy absorption into the measurable device parameters, such as current and voltage.

With direct detection design and high atomic number and electron density of CdTe, efficiencies higher than currently popular commercial phosphor/amorphous silicon-based detectors can be achieved when no external biasing is applied.

Theoretical analysis and experimental verifications provided in this work ascertain that the large area thin-film CdTe-based detector system has a high potential for implementation as an efficient low-cost, long life detector, suitable for clinical diagnostic imaging as well as with high-energy x-rays used in clinical radiation therapy.

References

1. Abbene, L., & Del Sordo, S. (2014). CdTe detectors. In *Comprehensive biomedical physics* (p. 30). Elsevier.
2. Jambi, L. K., et al. (2015). Evaluation of XRI-UNO CdTe detector for nuclear medical imaging. *Journal of Instrumentation, 10*(06), P06012.
3. Bertolli, M. (2008). *Solar cell materials. Course: Solid State II*. Department of Physics, University of Tennessee.
4. Goswami, D. Y., & Kreith, F. (2007). *Handbook of energy efficiency and renewable energy*. Crc Press.
5. Luque, A., & Hegedus, S. (2011). *Handbook of photovoltaic science and engineering*. Wiley.
6. Del Sordo, S., et al. (2009). Progress in the development of CdTe and CdZnTe semiconductor radiation detectors for astrophysical and medical applications. *Sensors, 9*(5), 3491–3526.
7. Abbene, L., et al. (2010). High-rate x-ray spectroscopy in mammography with a CdTe detector: A digital pulse processing approach. *Medical Physics, 37*(12), 6147–6156.
8. Eisen, Y., Shor, A., & Mardor, I. (1999). CdTe and CdZnTe gamma ray detectors for medical and industrial imaging systems. *Nuclear Instruments and Methods in Physics Research Section A: Accelerators, Spectrometers, Detectors and Associated Equipment, 428*(1), 158–170.

9. Miyajima, S. (2003). Thin CdTe detector in diagnostic x-ray spectroscopy. *Medical Physics, 30*(5), 771–777.
10. Scheiber, C., & Giakos, G. C. (2001). Medical applications of CdTe and CdZnTe detectors. *Nuclear Instruments and Methods in Physics Research Section A: Accelerators, Spectrometers, Detectors and Associated Equipment, 458*(1–2), 12–25.
11. Milbrath, B. D., et al. (2008). Radiation detector materials: An overview. *Journal of Materials Research, 23*(10), 2561–2581.
12. Verger, L., et al. (1997). New developments in CdTe and CdZnTe detectors for X and γ-ray applications. *Journal of Electronic Materials, 26*(6), 738–744.
13. Lee, Y.-J., et al. (2013). Optimization of an ultra-high-resolution parallel-hole collimator for CdTe semiconductor SPECT system. *Journal of Instrumentation, 8*(01), C01044.
14. Wald, F. (1977). Applications of CdTe. A review. *Revue de Physique Appliquée, 12*(2), 277–290.
15. NIST ESTAR Database. http://physics.nist.gov//PhysRefData//Star//Text//ESTAR.html
16. Zhao, W., & Rowlands, J. (1997). Digital radiology using active matrix readout of amorphous selenium: Theoretical analysis of detective quantum efficiency. *Medical Physics, 24*(12), 1819–1833.
17. Zhao, W., et al. (1997). Digital radiology using active matrix readout of amorphous selenium: Construction and evaluation of a prototype real-time detector. *Medical Physics, 24*(12), 1834–1843.
18. Lee, D. L., et al. (1998). Improved imaging performance of a 14ʺ×17ʺ direct radiography system using a Se/TFT detector. In *Medical imaging 1998: Physics of medical imaging*. SPIE.
19. Tsukamoto, A., et al. (1999). Development and evaluation of a large-area selenium-based flat-panel detector for real-time radiography and fluoroscopy. In *Medical imaging 1999: Physics of medical imaging*. International Society for Optics and Photonics.
20. Antonuk, L., et al. (2000). Strategies to improve the signal and noise performance of active matrix, flat-panel imagers for diagnostic x-ray applications. *Medical Physics, 27*(2), 289–306.
21. Simon, M., et al. (2004). PbO as direct conversion x-ray detector material. In *Medical imaging 2004: Physics of medical imaging*. International Society for Optics and Photonics.
22. Ouimette, D. R., Nudelman, S., & Aikens, R. S. (1998). New large-area x-ray image sensor. In *Medical imaging 1998: Physics of medical imaging*. SPIE.
23. Adachi, S., et al. (2000). Experimental evaluation of a-Se and CdTe flat-panel x-ray detectors for digital radiography and fluoroscopy. In *Medical imaging 2000: Physics of medical imaging*. SPIE.
24. Tokuda, S., et al. (2001). Experimental evaluation of a novel CdZnTe flat-panel x-ray detector for digital radiography and fluoroscopy. In *Medical imaging 2001: Physics of medical imaging*. SPIE.
25. Tokuda, S., et al. (2002). Large-area deposition of a polycrystalline CdZnTe film and its applicability to x-ray panel detectors with superior sensitivity. In *Medical imaging 2002: Physics of medical imaging*. International Society for Optics and Photonics.
26. Tokuda, S., et al. (2003). Improvement of the temporal response and output uniformity of polycrystalline CdZnTe films for high-sensitivity X-ray imaging. In *Medical imaging 2003: Physics of medical imaging*. International Society for Optics and Photonics.
27. Choi, C.-W., et al. (2007). Comparison of compound semiconductor radiation films deposited by screen printing method. In *Medical imaging 2007: Physics of medical imaging*. SPIE.
28. Hoheisel, M., Giersch, J., & Bernhardt, P. (2004). Intrinsic spatial resolution of semiconductor X-ray detectors: A simulation study. *Nuclear Instruments and Methods in Physics Research Section A: Accelerators, Spectrometers, Detectors and Associated Equipment, 531*(1–2), 75–81.
29. Cunningham, I. A. (2000). Applied linear-systems theory. *Handbook of Medical Imaging, 1*, 79–159.
30. Team, M.C., MCNP-A general N-Particle Transport Code, Version 5. vol. I, Overview and theory, 2005.

31. Shvydka, D., Jin, X., & Parsai, E. I. (2013). Performance of large area thin-film CdTe detector in diagnostic x-ray imaging. *International Journal of Medical Physics, Clinical Engineering and Radiation Oncology, 2*(03), 98.
32. Giger, M. L., & Doi, K. (1984). Investigation of basic imaging properties in digital radiography. I. Modulation transfer function. *Medical Physics, 11*(3), 287–295.
33. Lee, D. L., et al. (1996). Discussion on resolution and dynamic range of Se-TFT direct digital radiographic detector. In *Medical imaging 1996: Physics of medical imaging*. SPIE.
34. Boone, J. M., et al. (1999). A Monte Carlo study of x-ray fluorescence in x-ray detectors. *Medical Physics, 26*(6), 905–916.
35. Riederer, S. J., Pelc, N. J., & Chesler, D. A. (1978). The noise power spectrum in computed X-ray tomography. *Physics in Medicine & Biology, 23*(3), 446.
36. Boedeker, K. L., Cooper, V. N., & McNitt-Gray, M. F. (2007). Application of the noise power spectrum in modern diagnostic MDCT: Part I. Measurement of noise power spectra and noise equivalent quanta. *Physics in Medicine & Biology, 52*(14), 4027.
37. Cunningham, I. (2000). In J. Beutel, H. Kundel, & R. L. Van Metter (Eds.), *Handbook of medical imaging* (Vol. 1, p. 79). SPIE.
38. Swank, R. K. (1973). Absorption and noise in x-ray phosphors. *Journal of Applied Physics, 44*(9), 4199–4203.
39. Jaffray, D., et al. (1995). Monte Carlo studies of x-ray energy absorption and quantum noise in megavoltage transmission radiography. *Medical Physics, 22*(7), 1077–1088.
40. Fahrenbruch, A., & Bube, R. H. (1983). *Fundamentals of solar cells*. London Academic press Inc from Usenet.
41. Harju, R., et al. (2000). Electron-beam induced degradation in CdTe photovoltaics. *Journal of Applied Physics, 88*(4), 1794–1801.
42. Procz, S., et al. (2020). Investigation of CdTe, GaAs, Se and Si as sensor materials for mammography. *IEEE Transactions on Medical Imaging, 39*(12), 3766–3778.
43. Antonuk, L. E. (2002). Electronic portal imaging devices: A review and historical perspective of contemporary technologies and research. *Physics in Medicine & Biology, 47*(6), R31.
44. Tada, H., et al., Solar cell radiation handbook. 1982.
45. Sawant, A., et al. (2005). Segmented crystalline scintillators: An initial investigation of high quantum efficiency detectors for megavoltage x-ray imaging. *Medical Physics, 32*(10), 3067–3083.
46. Mei, X., Rowlands, J., & Pang, G. (2006). Electronic portal imaging based on Cerenkov radiation: A new approach and its feasibility. *Medical Physics, 33*(11), 4258–4270.
47. Pang, G., & Rowlands, J. (2004). Development of high quantum efficiency, flat panel, thick detectors for megavoltage x-ray imaging: A novel direct-conversion design and its feasibility: High quantum efficiency flat panel detectors. *Medical Physics, 31*(11), 3004–3016.
48. Zhao, W., & Rowlands, J. A. (1995). X-ray imaging using amorphous selenium: Feasibility of a flat panel self-scanned detector for digital radiology. *Medical Physics, 22*(10), 1595–1604.
49. Lachaine, M., & Fallone, B. (1998). Monte Carlo detective quantum efficiency and scatter studies of a metal/-Se portal detector. *Medical Physics, 25*(7), 1186–1194.
50. Zhao, W., et al. (2003). Imaging performance of amorphous selenium based flat-panel detectors for digital mammography: Characterization of a small area prototype detector. *Medical Physics, 30*(2), 254–263.
51. Kang, Y., et al. (2005). Examination of PbI/sub 2/and HgI/sub 2/photoconductive materials for direct detection, active matrix, flat-panel imagers for diagnostic X-ray imaging. *IEEE Transactions on Nuclear Science, 52*(1), 38–45.
52. Su, Z., et al. (2005). Systematic investigation of the signal properties of polycrystalline HgI2 detectors under mammographic, radiographic, fluoroscopic and radiotherapy irradiation conditions. *Physics in Medicine & Biology, 50*(12), 2907.
53. McCandless, B. E., & Sites, J. R. (2003). Cadmium telluride solar cells. In *Handbook of photovoltaic science and engineering* (pp. 617–662). Wiley.
54. Booth, T. (2003). *A general Monte Carlo N-particle transport code, version 5, volume 1: Overview and theory*. Los Alamos National Laboratory.

55. Kang, J., et al. (2008). From photovoltaics to medical imaging: Applications of thin-film CdTe in x-ray detection. *Applied Physics Letters, 93*(22), 223507.
56. Antonuk, L. E., et al. (1998). Initial performance evaluation of an indirect-detection, active matrix flat-panel imager (AMFPI) prototype for megavoltage imaging. *International Journal of Radiation Oncology* Biology* Physics, 42*(2), 437–454.
57. Burgelman, M., Nollet, P., & Degrave, S. (2000). Modelling polycrystalline semiconductor solar cells. *Thin Solid Films, 361*, 527–532.
58. Parsai, E. I., Shvydka, D., & Kang, J. (2010). Design and optimization of large area thin-film CdTe detector for radiation therapy imaging applications. *Medical Physics, 37*(8), 3980–3994.
59. Wowk, B., et al. (1994). Optimization of metal/phosphor screens for on-line portal imaging. *Medical Physics, 21*(2), 227–235.
60. Whelpton, D., & Watson, B. (1963). A pn junction photovoltaic detector for use in radiotherapy. *Physics in Medicine & Biology, 8*(1), 33.
61. Scharf, K. (1960). Photovoltaic effect produced in silicon solar cells by X-and gamma rays. *Journal of Research of the National Bureau of Standards. Section A, Physics and Chemistry, 64*(4), 297.
62. Fahrenbruch, A., & Bube, R. (2012). *Fundamentals of solar cells: photovoltaic solar energy conversion*. Elsevier.
63. Shvydka, D., Karpov, V., & Compaan, A. (2002). Bias-dependent photoluminescence in CdTe photovoltaics. *Applied Physics Letters, 80*(17), 3114–3116.
64. Akbari, F., & Shvydka, D. (2022). Electron backscattering for signal enhancement in a thin-film CdTe radiation detector. *Medical Physics, 49*, 6654–6665.
65. Orita, T., et al. (2021). Double-photon emission imaging with high-resolution Si/CdTe Compton cameras. *IEEE Transactions on Nuclear Science, 68*(8), 2279–2285.
66. Sakai, M., et al. (2019). Compton imaging with 99mTc for human imaging. *Scientific Reports, 9*(1), 1–8.
67. Nakano, T., et al. (2020). Imaging of 99mTc-DMSA and 18F-FDG in humans using a Si/CdTe Compton camera. *Physics in Medicine & Biology, 65*(5), 05LT01.
68. Shvydka, D., et al. (2013). WE-G-500-08: Novel multilayer detector design using poly-crystalline CdTe for radiation therapy imaging applications. *Medical Physics, 40*(6Part30), 504–504.
69. Guanggen, Z., et al. (2013). The effect of irradiation on the mechanism of charge transport of CdTe solar cell. In *2013 IEEE 39th Photovoltaic Specialists Conference (PVSC)*. IEEE.
70. Loferski, J. J. (1966). The effects of electron and proton irradiation on thin film solar cells. *Revue de Physique Appliquee, 1*(3), 221–227.
71. Zanarini, M., et al. (2004). Radiation damage induced by 2 MeV protons in CdTe and CdZnTe semiconductor detectors. *Nuclear Instruments and Methods in Physics Research Section B: Beam Interactions with Materials and Atoms, 213*, 315–320.
72. Cho, S., et al. (2019). Radiation hardness of cadmium telluride solar cells in proton therapy beam mode. *PLoS One, 14*(9), e0221655.
73. Miyamaru, H., et al. (1997). Effect of fast neutron irradiation on CdTe radiation detectors. *Journal of Nuclear Science and Technology, 34*(8), 755–759.
74. Chester, R. O. (1967). Radiation damage in cadmium sulfide and cadmium telluride. *Journal of Applied Physics, 38*(4), 1745–1752.
75. Attlx, F. H. (1986). *Introduction to radiological physics and radiation dosimetry*. Wiley-VCH.
76. Konobeyev, A. Y., et al. (2017). Evaluation of effective threshold displacement energies and other data required for the calculation of advanced atomic displacement cross-sections. *Nuclear Energy and Technology, 3*(3), 169–175.
77. Shvydka, D., Parsai, E., & Kang, J. (2008). Radiation hardness studies of CdTe thin films for clinical high-energy photon beam detectors. *Nuclear Instruments and Methods in Physics Research Section A: Accelerators, Spectrometers, Detectors and Associated Equipment, 586*(2), 169–173.
78. Bonnet, D. (2012). CdTe thin-film PV modules. In *Practical handbook of photovoltaics* (pp. 283–322). Elsevier.

79. Corkish, R. (2013). Solar cells. In *Reference module in earth systems and environmental sciences* (p. 15). Elsevier.
80. Shvydka, D., et al. (2007). TH-C-M100E-03: A new generation of Electronic Portal Imaging Devices (EPID) using thin-film CdTe for radiation oncology applications. *Medical Physics, 34*(6Part23), 2629–2629.
81. Afaneh, F., et al. (2018). The γ-irradiation effect on the optical properties of CdTe thin films deposited by thermal evaporation technique. *Materials Science, 24*(1), 3–9.
82. Okamoto, T., et al. (2021). Gamma-ray irradiation effects on CdTe solar cell dosimeter. *Japanese Journal of Applied Physics, 60*(SB), SBBF02.
83. Redfield, D., & Bube, R. H. (1996). *Photo-induced defects in semiconductors*. Cambridge University Press.

Investigation of Structural Defects of (Cd, Zn)Te Crystals Grown by the Traveling Heater Method

Jiaona Zou, Alex Fauler, and Michael Fiederle

1 Introduction

Radiation detectors are widely applied for national security, nonproliferation inspections, medical imaging, space exploration, and astrophysics investigation [1]. These applications demand radiation detectors with high resolution, low noise, and high detection efficiency. (Cd, Zn)Te (CZT) crystals have proven to be a suitable candidate for developing such radiation detectors due to their electrical properties, room temperature operability, and absorption efficiency for X-rays. Detector-grade CZT crystals were initially grown using the Bridgman method. In recent years, the traveling heater method (THM) is demonstrated to be an excellent growth technique for producing large-volume detector-grade CZT crystals with high uniformity by avoiding Zn segregation [2–7]. Nevertheless, CZT is susceptible to deformation by stress generated during crystal growth because of its high ionicity and low yield stress [8]. Moreover, the performance, uniformity, and efficiency of detector devices and the processability of crystals are still limited by compositional inhomogeneities and structural defects, such as Te inclusions, dislocations, grain/subgrain boundaries, twins, and even cracks in the material [9–12]. Twins and grain boundaries form potential barriers for the drifting carriers and cause electrical diffusion, affecting charge transport. The associated dislocations can further enhance charge trapping by accumulating secondary phases and impurities along the boundaries [13, 14]. Te inclusions can also trap the charges from the electron clouds. Because of their random distribution of the interaction points, there are variations in trapped charges, resulting in significant fluctuations in the collected charge signal in the detector device [15]. Additionally, the magnitude of the trapping

J. Zou (✉) · A. Fauler · M. Fiederle
Freiburg Materials Research Center, Freiburg, Germany
e-mail: jiaona.zou@fmf.uni-freiburg.de

© The Author(s), under exclusive license to Springer Nature Switzerland AG 2023   43
L. Abbene, K. (Kris) Iniewski (eds.), *High-Z Materials for X-ray Detection*,
https://doi.org/10.1007/978-3-031-20955-0_3

effect of inclusions depends strongly on their size [13]. It is therefore imperative to understand the influence of growth conditions on defect formation and growth mechanisms to enhance the availability of large-volume CZT crystals.

This chapter concentrates primarily on the crystal growth of CZT by the THM and the characterization of compositional uniformity and structural defects in as-grown CZT crystals. In the following text, the material requirements for radiation detector devices and the THM are described. As examples, we analyzed two $Cd_{0.9}Zn_{0.1}Te$ crystals grown by the THM under the rotating magnetic field (RMF). The compositional uniformity, Te inclusions, the stress associated with defects, and the resistivity were explored.

2   Detector Material Requirements

There are some important material requirements of a semiconductor when selected and used as a radiation detector capable of operating at room temperature. These requirements consist of a large energy bandgap E_g, a high atomic number Z, a high mobility-lifetime product, a high resistivity, and a high homogeneity of the material with low density of defects:

(a) A large bandgap (>1.4 eV) is required to prevent the thermal generation of carriers and thus ensure a low dark current.
(b) The importance of a high atomic number Z is to enhance the probability of the radiation interaction.
(c) To sustain a large electric field and ensure a low dark current, a high resistivity ($>10^7$ Ωcm) is necessary.
(d) A high mobility-lifetime product is required to ensure a long carrier drift length and allow for charge transport and collection.
(e) A homogeneous material with a low concentration of defects is of great importance for the fabrication of large volume detectors with high efficiency [16].

Among the most commonly used semiconductor detector materials, CZT has attracted a lot of interest due to its outstanding properties. Because of the high atomic number and the high density, CZT detectors ensure the radiation interaction coefficient and thus the detection efficiency. Furthermore, CZT has a high mobility-lifetime product, so the charge carriers do not recombine before being collected. Most importantly, a CZT detector can be operated at room temperature due to its wide bandgap. These properties together make CZT a very desirable and potential detector material.

However, the commercial potential of CZT used for high-energy radiation detection is underutilized due to the poor uniformity and crystallinity of the crystals. Crystal defects, such as inclusions, grain boundaries, twins, and even cracks, can dramatically affect the detector performance [9].

3 Traveling Heater Method

The application of CZT materials as radiation detectors demands high CZT crystal quality, a large volume at acceptable costs. However, the growth of such good-quality and large-volume crystals remains a challenge. For the last few decades, the most popular technique for growing bulk crystals is the growth from congruent or near-congruent melts. The most common methods for bulk CdTe and CZT crystals are the high pressure Bridgman, the conventional vertical Bridgman, and the horizontal Bridgman methods. There are some drawbacks of these Bridgman methods. First, the growth is undertaken at high temperatures near the congruent melting point of the material. The high growth temperature and reactions with the crucibles cause the formation of defects and contaminations. Second, the formation of Te-rich secondary phases during the growth or cooling process severely affects the detector's performance. Last, the segregation of Zn leads to inhomogeneity of the mixed crystal [17, 18].

The THM has received much attention due to the successful production of large quantities of detector-grade CdTe and CZT crystals [5, 6, 19]. The THM combines zone melting and growth from solutions. The configuration of the THM is rather similar to liquid phase epitaxy. A CZT seed crystal is placed at the bottom of the ampoule, which serves as a substrate. A Te-rich solvent zone, located between the seed and a polycrystalline CZT source material or so-called feed material, is melted by a narrow heater. During growth, a slow translation of the ampoule relative to the heater is accomplished. Consequently, the CZT feed material is dissolved at the hot upper interface of the solvent zone, and a new CZT material cools down and deposits at the cool lower interface. Convection and diffusion, driven by the temperature gradient, are the dominant mechanisms of mass transport from the feed material to the growing crystal [20]. Gradually as the solvent zone moves, the CZT feed crystal dissolves and subsequently deposits on the seed crystal.

There are several advantages of the THM for the growth of CZT crystals. The growth occurs from the Te-rich solvent zone, indicating that the growth temperature can be 300–350 °C lower than the melting point of CZT. The resulting benefits are fewer contaminations from the container, lower thermal stress, and fewer intrinsic defects such as dislocations and Te precipitates. Another advantage is the precise and uniform composition of the regrown crystal. Higher purity of the CZT crystals with axial and radial compositional uniformity and fewer contaminations were produced by the THM when compared to those from Bridgman [21]. As the feed material is gradually dissolved, the solution zone will reach saturation, and extra material with nearly the same composition will be deposited at the growth front. In addition, the segregation coefficients of many impurities, such as Li, Ti, Fe, Cu, Ag, Au, and so on, are lower in Te solvent than in CdTe and CZT [22]. Consequently, the impurities tend to segregate more efficiently in Te-rich zone than in the growing CZT crystal. Furthermore, the retrograde Te solubility with temperature [23] can result in a smaller size, and lower density of Te precipitates in CZT crystals grown from the THM than melt-growth techniques.

Despite the aforementioned advantages, the THM has some disadvantages making it still challenging to produce large homogeneous crystals. Since the growth occurs from the Te-rich solvent zone, the Te-rich solution can be entrapped into the regrown crystals due to any microscopic irregularity, forming as a result a high concentration of Te-rich secondary phases [4]. In addition, the significantly lower growth rate, ranging from 1 to 7.5 mm/day, remains a severe problem in the industrial growth by THM. Furthermore, a seed crystal, most commonly single crystalline, is required as a substrate. Finally, the uncontrolled convectional flow [24] and the temperature gradient at the growth interface affect the stability and shape of the growth interface, which in turn influences the crystal quality and grain size [4, 22].

4   Compositional Uniformity

In the alloy $Cd_{1-x}Zn_xTe$, the fraction x of Cd atoms is considered to be randomly substituted by the Zn atoms. The Zn atoms are smaller than Cd atoms, and the Ze-Te bond length and ionicity are smaller than those of Cd-Te. The incorporation of Zn thus strengthens the CdTe lattice and increases the hardness, resulting in increased material stability of $Cd_{1-x}Zn_xTe$ alloys. Consequently, the density of dislocations and subgrains is reduced [22].

In crystal growth from solution such as the THM, the composition of the growing crystal depends on the solution composition ahead of the crystallization front, especially on the ZnTe/CdTe ratio, and the uniformity of the feed material. If the composition in the solvent is not consistent with that in the feed, the ZnTe composition tends to vary as the growth continues. The resulting composition variation in ZnTe with position further leads to a variation in the bandgap, the number of photogenerated electron-hole pairs, and the leakage current in the detection device [25]. Therefore, it is of great importance to have a uniform ZnTe concentration along the crystal for device fabrication.

More homogeneous ZnTe content can be obtained by adjusting the composition of the Te-rich solution zone in a THM system. Figure 1 illustrates the ZnTe content distributions along the growth direction of two $Cd_{0.9}Zn_{0.1}Te$ crystals grown by the THM. In these specific examples, we can easily find distinct ZnTe content variations at the interface for both crystals. The ZnTe variation at the interface of the crystal F1-02 is 2.51% mole fraction, while that of another crystal F1-03 is about 1.02%. The efficient reduction of ZnTe content in F1-03 was achieved by enhancing the CdTe composition in the Te-rich solution zone from 5% to 10%. Furthermore, the ZnTe concentration of both crystals decreases gradually as the growth/solidification continues and approaches the value of the ZnTe content in the feeds at the end of growth, which is 10% mole fraction. Since the ZnTe concentration in the crystal at beginning of growth is higher than that in the feed material, the amount of ZnTe outgoing from the solution into the growing crystal exceeds the ZnTe amount

Fig. 1 ZnTe content distributions along the growth direction of the crystals and calculated value from mathematical simulation [21]. ZnTe content variation at the interface of F1-02 is higher than that in F1-03. In both crystals, gradual reductions in the ZnTe content are observed as the growth/solidification continues [26]

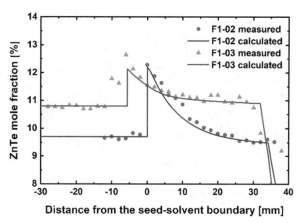

received from the feed. The solution is thus gradually depleted with ZnTe, resulting in decreasing ZnTe content in the growing crystal.

5 Inclusions

Since Te-rich solution is used in the THM growth, the emergence of Te as a second phase in solidified crystals from the mother liquid is inevitable. The second phase Te presents itself into crystalline CZT in two forms: as precipitates and inclusions. Firstly, because the solid solubility of CdTe is retrograde, extra Cd or Te will precipitate on cooling in the form of point defects. When the composition of the melt is Te-saturated, Te precipitates will form. At the growth under near stoichiometric conditions, their density can be effectively minimized [27]. Te precipitates behave like point defects in CZT crystals.

Unlike precipitates, inclusions are formed during the melt growth of crystals by capturing melt-solution droplets from the diffusion boundary at the growing interface's front [27]. They preferably concentrate at the re-entrant angles of the grain boundaries and twins crossing the growth interface. Since CZT crystals are transparent to infrared light whereas Te inclusions are opaque, Te inclusions can be visualized by infrared transmission microscopy. The diameter of typical Te inclusions is 1–30 μm. Larger Te inclusions, up to a few hundred micrometers in diameter, were found in high-pressure, vertical-Bridgman, and THM-grown CZT. It was reported that Te inclusions with a diameter larger than 10 μm can dramatically affect the performance and efficiency of the detector, while the detector's spectral resolution is better when the inclusions' size is smaller [28, 29], and smaller size inclusions (<3 μm) show no effect on charge transport [30]. In addition, the peak width (%FWHM) of the response function of detectors narrows from 3.7% to 0.3% when the concentration of large Te inclusions (>10 μm) is reduced by two orders of

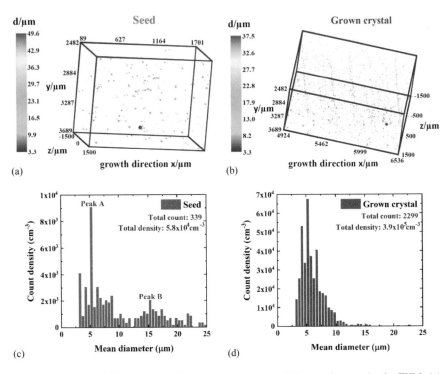

Fig. 2 Magnified 3D illustrations of the two regions from a CZT crystal grown by the THM: (**a**) seed and (**b**) the grown crystal. The inclusions randomly decorate the seed, while distinct striations are formed in the grown crystal. Histogram of the inclusion size distribution in the seed (**c**) and the grown crystal (**d**). Almost 7 times more inclusions are found in the grown crystal (from 339 to 2299). However, the average size of the inclusions is reduced to half (from 10.57 to 6.17 μm) of that in the seed [32]

magnitude [28]. Therefore, regarding applications as radiation detectors, efforts to increase the detector efficiency have focused on reducing the size of the inclusions.

By using the THM growth, a reduction of almost half of the Te inclusion size can be achieved. Figure 1 demonstrates a comparison of the 3D distribution of Te inclusions in two regions with the same size from the seed and the grown crystal by the THM. The original seed with a composition of $Cd_{0.9}Zn_{0.1}Te$ was grown using the Bridgman method [31]. In the Bridgman seed, some large inclusions are randomly distributed, while there are more inclusions in the THM grown crystal with considerably smaller sizes and distinct striation-like patterns. The number of Te inclusions in the grown crystal shows a sevenfold increase (from 339 to 2299), and their mean size almost halves (from 10.57 to 6.17 μm) compared to those in the seed. Furthermore, two peaks appear in the inclusion distribution histogram in the seed, as shown in Fig. 2c. Although the count concentration of Peak A is higher than that of Peak B, Peak B accounts for the majority of the inclusion volume due to their larger sizes. This typical bimodal or multimodal distribution in CZT/CdTe

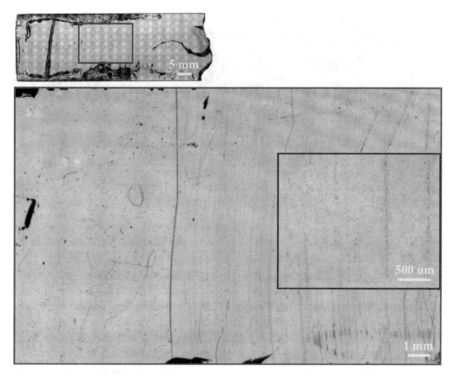

Fig. 3 Randomly distributed inclusions in the seed and striation-like patterns in the grown crystal. The periodicity of these inclusion striations is about 150–750 μm [32]

was observed earlier [33, 34]. However, this distribution is not found in the grown crystal region. In other words, the concentration of large Te inclusions (>10 μm) is significantly reduced in the THM-grown crystal than in the Bridgman seed.

It is evident that, in both Figs. 2 and 3, the inclusion distribution in the Bridgman seeds is random, while distinct striation-like patterns are displayed in the THM-grown crystals. This interesting phenomenon is sketched and illustrated in Fig. 4. The shape of these inclusion striations reveals the concave shape of the growth interface. The periodicity of these inclusion striations ranges from 150 to 750 μm. Since the translation rate is 0.15 mm/h, the periodicity of these inclusion striations is about 1–5 h of growth.

Striations are usually observed in silicon crystals grown using the Czochralski method [35]. Striations can vary in nature. For example, resistivity striations in Ge [36] and InSb [37] and carrier lifetime striations in silicon solar cells [38] were reported. Most of these striations arise from the rotation of the growing crystal, or from time-dependent flow phenomena in the melt, such as a fluctuating growth rate. Te inclusion striations in CdTe with periodicity in the range of several centimeters were first observed by Bolke et al. [28] in 2017.

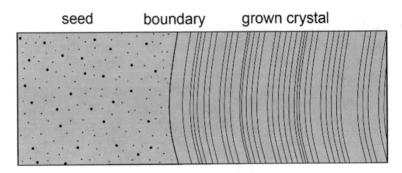

Fig. 4 Schematic striation-like patterns of the Te inclusions in THM grown crystals with a concave interface [32]

The formation of Te inclusions in a grown crystal was demonstrated to be strongly dependent on the microscopic morphology of the growth interface [38], which in turn can be influenced by growth parameters such as rotation, growth rate, growth temperature, temperature gradient, and temperature stability of the furnace. In this case, these inclusion striations are related to periodic temperature variations. The periods of inclusion distribution along the growth direction in a specific area and the corresponding ampoule temperature during this growth period are depicted in Fig. 5. The most prevalent period of the temperature variations is 1.57 h, which matches well with the most salient period for inclusion distribution of 254 μm. This leads to a small difference of 19 μm, which can be attributed to the error in data acquisition and the changing local growth rate. This suggests that the axial distribution of Te inclusions in this crystal tends to synchronize with the corresponding periodic temperature variations. That is, the inclusion striations are caused by the periodic temperature variations during growth. It should be emphasized that it is not the entirety of the temperature fluctuations that contributes to the inclusion striation-like patterns, but rather the low-frequency periodic components of the temperature variations. The explanation is as follows. The periodic temperature variations result in fluctuations in the instantaneous growth rate. According to the inclusion capture theory of Chernov and Temkin [39], the particles or droplets of a secondary phase will be captured by the solid/melt interface when the growth rate ϑ is above a critical growth rate ϑ_c. Conversely, the particles will be repelled and pushed forward by the interface if ϑ_c is not reached. According to this theory, fluctuations in the growth rate will cause variations in the distribution and size of the inclusions formed at the growth interface.

6 Defects and Stress

Defects are expected to be associated with stress, which can be evaluated by birefringence and X-ray white beam topography (XWBT). Birefringence is a

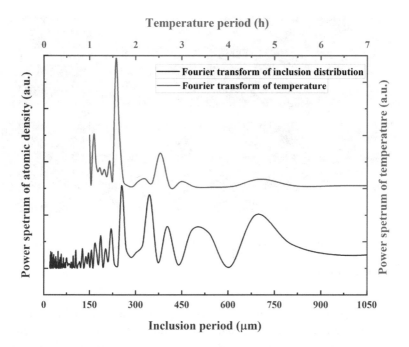

Fig. 5 Fourier transform results of the axial distribution of Te inclusions and the temporal change in the temperature. The synchronization indicates that the inclusion striations are created by the periodic temperature fluctuations [32]

phenomenon in which light travels into two rays with different phases when it enters a crystallographic non-equivalent axis in an anisotropic crystal. In isotropic crystals, which have equivalent axes, the interactions with light occur in a similar manner. In other words, the phenomenon of birefringence will not be observed in isotropic crystals, such as CZT with cubic structure. However, if the isotropic crystals have some defective or strained areas creating optical anisotropic regions, the phase of the light will be shifted. Since the amount of the phase shift is dependent on the inner stress, the inner stress inside the crystals can be estimated with a 2D image.

In Fig. 6, some typical defects in CZT crystals and their related local stress fields are clearly illustrated in the infrared transmission images and birefringence images, respectively. Grain boundaries and twins can generate different degrees of stress fields. For instance, a grain boundary in Fig. 6a, b induces strong stress; however, the grain boundaries in Fig. 6c, d show different degrees of birefringence contrast. The twin boundary in Fig. 6e, f generates a rather different stress field when compared to the grain boundaries. This suggests that the stress associated with the wall contact is relieved by the formation of twins. Inclusions are often associated with other defects in the lattice and nucleate at planer boundaries, such as grain and twin boundaries, as shown in Fig. 6a, c, e. These inclusions are relatively small, with a size of a few micrometers. In these regions, no stress is generated by the inclusions themselves, as

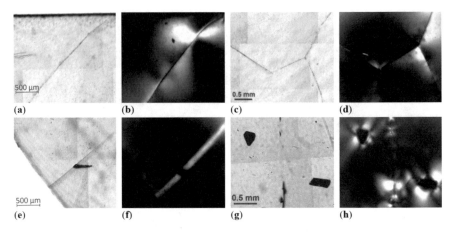

Fig. 6 Infrared transmission images and birefringence images showing related local stress fields of typical defects observed in CZT crystals: grain boundaries (**a–d**); a twin boundary (**e**) and (**f**); Te inclusions (**g**) and (**h**) [26]

shown in the birefringence maps. Two large inclusions with a size of a few hundred micrometers are illustrated in Fig. 6g, h, one with a hexagonal or near triangular shape and the other one with an elongated rectangular shape. In addition, some pearl string inclusions present themselves in the middle part of the image. All of these inclusions induce relatively strong stress fields. The Te droplet entrapment of a star-like shape corresponds to the local stress generation at the hexagon and rectangle edges [22].

Besides birefringence, X-ray topography is another powerful technique used for visualization, by the means of X-ray diffraction, of microstructural defects in crystals, such as precipitates, voids, dislocations, stacking faults, strained layers, and surface damage. When the crystal is illuminated by an X-ray beam, the beam will be diffracted in many directions according to Bragg's law, as a result creating Laue spots. If the crystal is a perfect crystal, each spot contains a homogeneous intensity. If, on the other hand, the crystal is strained, the deviations in diffracted intensity and interference of diffracted beams due to variations in lattice spacing Δd, in turn, result in the changes in the image contrast and formation of streaks. Consequently, the lattice deformations surrounding the defects are seen even though defects themselves are not visible in the image.

As an example, back reflection topographs recorded from the interface area of a CZT crystal grown by the THM are shown in Fig. 7a, and the corresponding infrared image of the scanned region is shown in Fig. 7b. Excessive stress with white contrast in the peripheral regions of the slabs is observed, which originates from the ampoule wall and the huge inclusions (TI). These inclusions are formed due to the use of the Te-rich solution from the THM. Scratches (Sc) appear on the surface of both sides. It is evident that one large inclusion (TI1) in the seed generates high degrees of stress, shown as white contrast in the XWBT image. Several comparatively smaller

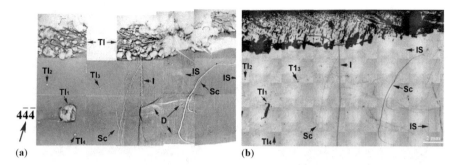

Fig. 7 XWBT back reflection mapping (**a**) and (**b**) infrared image (**b**) near the interface region of a CZT crystal grown by the THM. Some scratches (Sc) exist on the surface. The observation of white dots suggests the stress induced by Te inclusions (TI1–TI4). Extensive distortion and dislocation networks (D) accumulate at the interface and further extend into the grown crystal region. White streaks prevail the grown crystal, which are considered to be caused by the inclusion striations (IS) [26]

inclusions (TI2–TI4), displayed as smaller white dots, can be found in the Bridgman seed. In contrast, no inclusions are visible in the mappings of the THM-grown crystal. The infrared image also confirms the result that the inclusion size in the grown crystal is much smaller than in the seed. It is worth noting that in the seed interface region (I), high lattice distortion is observed, which is mainly generated by the ZnTe concentration variation at the interface, as discussed in Sect. 4. The stress at the interface propagates forward into the grown crystal region, giving rise to the formation of dislocation networks (D). Additionally, inclusion striations (IS) are slightly visible as white streaks, and they prevail the whole grown crystal, showing a concave shape.

7   Resistivity

High bulk resistivity, wide bandgap, and high charge transport capability are essential properties for CZT crystals used as room temperature radiation detectors. The CZT semiconductor has a direct wide energy bandgap in the range of 1.4–2.2 eV, which is strongly dependent on the Cd:Zn ratio and temperature according to the study of Tobin [29]. The resistivity of a CZT semiconductor increases exponentially with its bandgap, which is related to Zn composition. Thus, the intrinsic resistivity of $Cd_{1-x}Zn_xTe$ increases with increasing Zn concentration. Figure 8 illustrates the resistivity of a $Cd_{0.9}Zn_{0.1}Te$ crystal from a contactless resistivity mapping. The resistivity is in the order of 3×10^9 Ωcm with high homogeneity across the whole slab, which is consistent with high-quality CZT materials [40–42]. Several regions in the grown crystal exhibit relatively high resistivity than that of other regions. It is assumed that the abundant grain boundary defects and the cracks including the associated dislocation networks are related to the increase in resistivity. Such defects

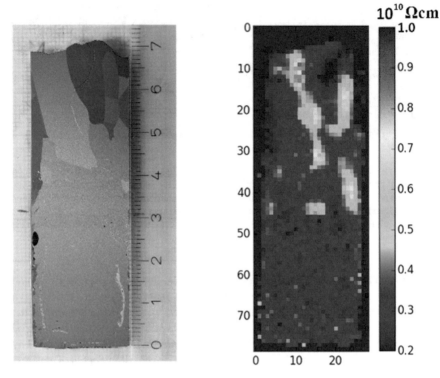

Fig. 8 Contactless resistivity mapping of a CZT crystal [26]

affect the charge carrier density by trapping free charge carriers, which further leads to resistivity variations.

8 Conclusion and Outlook

CZT is a promising material for radiation detection. The crystal growth by the modern growth technique of the THM yields remarkable results for large-volume high-quality CZT crystals. However, material defects remain the limiting factors for the development of radiation detectors. In this chapter, the structural defects of CZT crystals grown by the THM were investigated. These CZT crystals exhibited highly uniform composition and good optical and electrical performance. The ZnTe content was successfully controlled by adjusting the composition in the Te-rich solvent zone. In addition, the size of the Te inclusions in the THM-grown crystal demonstrated a distinct reduction than that in the Bridgman seed. Inclusion striations were observed, which were found to be induced by the periodic temperature fluctuations. Furthermore, the CZT crystals showed a high resistivity of 3×10^9 Ωcm.

To further improve the crystal quality, the growth mechanisms and defect formation need to be better understood. Since defects and inhomogeneity are strongly related to gravitational convection, microgravity can be used as a tool to advance this knowledge and improve the crystal growth on Earth [43]. A series of long-term CZT growth experiments by the THM under microgravity, dubbed "VAMPIR-F," are scheduled and will be carried out in the Russian Multifunctional Laboratory Module (MLM) on the International Space Station (ISS). These THM experiments under microgravity will provide optimal conditions to understand diffusion during the growth process and defect formation and will be of great scientific significance for the improvement of crystal growth in the laboratory and industrial processes in the near future.

References

1. Yang, G., & James, R. B. (Eds.). (2010). Chapter IIC – applications of CdTe, CdZnTe, and CdMnTe radiation detectors. In *CdTe and related compounds; Physics, defects, hetero- and nano-structures, crystal growth, surfaces and applications* (pp. 145–255). Elsevier. https://doi.org/10.1016/B978-0-08-096513-0.00002-9

2. Zhou, B., et al. (2018, February). Modification of growth interface of CdZnTe crystals in THM process by ACRT. *Journal of Crystal Growth, 483*, 281–284. https://doi.org/10.1016/j.jcrysgro.2017.12.003

3. Chen, H., et al. (2018, September). Development of large-volume high-performance monolithic CZT radiation detector. In *Hard X-ray, gamma-ray, and neutron detector physics XX* (Vol. 10762, p. 107620N). https://doi.org/10.1117/12.2321244

4. Roy, U. N., Burger, A., & James, R. B. (2013, September). Growth of CdZnTe crystals by the traveling heater method. *Journal of Crystal Growth, 379*, 57–62. https://doi.org/10.1016/j.jcrysgro.2012.11.047

5. Shiraki, H., Funaki, M., Ando, Y., Tachibana, A., Kominami, S., & Ohno, R. (2009, August). THM growth and characterization of 100 mm diameter CdTe single crystals. *IEEE Transactions on Nuclear Science, 56*(4), 1717–1723. https://doi.org/10.1109/TNS.2009.2016843

6. Chen, H., et al. (2008, January). Characterization of large cadmium zinc telluride crystals grown by traveling heater method. *Journal of Applied Physics, 103*(1), 014903. https://doi.org/10.1063/1.2828170

7. Iniewski, K. (2016). CZT sensors for computed tomography: From crystal growth to image quality. *Journal of Instrumentation, 11*(12), C12034–C12034. https://doi.org/10.1088/1748-0221/11/12/C12034

8. Chung, H., Raghothamachar, B., Dudley, M., & Larson, D. J., Jr. (1996, July). Synchrotron white beam X-ray topography characterization of structural defects in microgravity and ground-based CdZnTe crystals. In *Space processing of materials* (Vol. 2809, pp. 45–56). https://doi.org/10.1117/12.244357

9. Bolotnikov, A. E., et al. (2011, August). Correlations between crystal defects and performance of CdZnTe detectors. *IEEE Transactions on Nuclear Science, 58*(4), 1972–1980. https://doi.org/10.1109/TNS.2011.2160283

10. Hossain, A., et al. (2017, July). Direct observation of influence of secondary-phase defects on CZT detector response. *Journal of Crystal Growth, 470*, 99–103. https://doi.org/10.1016/j.jcrysgro.2017.04.002

11. Buis, C., Marrakchi, G., Lafford, T. A., Brambilla, A., Verger, L., & Gros d'Aillon, E. (2013, February). Effects of dislocation walls on image quality when using cadmium telluride X-ray detectors. *IEEE Transactions on Nuclear Science, 60*(1), 199–203. https://doi.org/10.1109/TNS.2012.2232306

12. Triboulet, R., & Siffert, P. (Eds.). (2010). Chapter V – Defects. In *CdTe and related compounds; physics, defects, hetero- and nano-structures, crystal growth, surfaces and applications* (pp. 169–307). Elsevier. https://doi.org/10.1016/B978-0-08-046409-1.00005-8

13. Bolotnikov, A. E., et al. (Aug. 2009). Extended defects in CdZnTe radiation detectors. *IEEE Transactions on Nuclear Science, 56*(4), 1775–1783. https://doi.org/10.1109/TNS.2009.2019960

14. Camarda, G. S., et al. (Oct. 2011). Effect of extended defects in planar and pixelated CdZnTe detectors. *Nuclear instruments and methods in physics research section A: Accelerators, spectrometers, detectors and associated equipment, 652*(1), 170–173. https://doi.org/10.1016/j.nima.2010.12.012

15. Bolotnikov, A. E., et al. (Apr. 2010). Te inclusions in CZT detectors: New method for correcting their adverse effects. *IEEE Transactions on Nuclear Science, 57*(2), 910–919. https://doi.org/10.1109/TNS.2010.2042617

16. Schieber, M., James, R. B., & Schlesinger, T. E. (1995). Chapter 15 – Summary and remaining issues for room temperature radiation spectrometers. In *Semiconductors and semimetals* (Vol. 43, pp. 561–583). Elsevier. https://doi.org/10.1016/S0080-8784(08)62754-4

17. Rudolph, P., & Mühlberg, M. (1993, January). Basic problems of vertical Bridgman growth of CdTe. *Materials Science and Engineering: B, 16*(1), 8–16. https://doi.org/10.1016/0921-5107(93)90005-8

18. Rudolph, P. (1994, January). Fundamental studies on Bridgman growth of CdTe. *Progress in Crystal Growth and Characterization of Materials, 29*(1), 275–381. https://doi.org/10.1016/0960-8974(94)90009-4

19. Amman, M., et al. (Jun. 2009). Evaluation of THM-grown CdZnTe material for large-volume gamma-ray detector applications. *IEEE Transactions on Nuclear Science, 56*(3), 795–799. https://doi.org/10.1109/TNS.2008.2010402

20. Triboulet, R. (2015). 12 – Crystal growth by traveling heater method. In P. Rudolph (Ed.), *Handbook of crystal growth* (2nd ed., pp. 459–504). Elsevier. https://doi.org/10.1016/B978-0-444-63303-3.00012-2

21. El Mokri, A., Triboulet, R., Lusson, A., Tromson-Carli, A., & Didier, G. (1994, April). Growth of large, high purity, low cost, uniform CdZnTe crystals by the 'cold travelling heater method. *Journal of Crystal Growth, 138*(1), 168–174. https://doi.org/10.1016/0022-0248(94)90800-1

22. Triboulet, R. (Ed.). (2010). Chapter IB – CdTe and CdZnTe growth. In *CdTe and related compounds; Physics, defects, hetero- and nano-structures, crystal growth, surfaces and applications* (pp. 1–144). Elsevier. https://doi.org/10.1016/B978-0-08-096513-0.00001-7

23. Steininger, J., Strauss, A. J., & Brebrick, R. F. (1970). Phase diagram of the Zn-Cd-Te ternary system – IOPscience. *The Electrochemical Society, 117*(10), 1305–1309.

24. Aggarwal, M. D., Batra, A. K., Lal, R. B., Penn, B. G., & Frazier, D. O. (2010). Bulk single crystals grown from solution on earth and in microgravity. In G. Dhanaraj, K. Byrappa, V. Prasad, & M. Dudley (Eds.), *Springer handbook of crystal growth* (pp. 559–598). Springer. https://doi.org/10.1007/978-3-540-74761-1_17

25. James, R. B., Schlesinger, T. E., Lund, J., & Schieber, M. (1995). Chapter 9 – Cd1-xZnxTe spectrometers for gamma and X-ray applications. In *Semiconductors and semimetals* (Vol. 43, pp. 335–381). Elsevier. https://doi.org/10.1016/S0080-8784(08)62748-9

26. Zou, J., et al. (2021 November). Characterization of structural defects in (Cd, Zn)Te crystals grown by the travelling heater method. *Crystals, 11*(11), 11. https://doi.org/10.3390/cryst11111402

27. Rudolph, P. (2003). Non-stoichiometry related defects at the melt growth of semiconductor compound crystals – A review. *Crystal Research and Technology, 38*(7–8), 542–554. https://doi.org/10.1002/crat.200310069

28. Bolke, J., et al. (2017, September). Measuring Te inclusion uniformity over large areas for CdTe/CZT imaging and spectrometry sensors. In *Sensors, systems, and next-generation satellites XXI* (Vol. 10423, p. 104231M). https://doi.org/10.1117/12.2278584

29. Tobin, S. P., et al. (1995, May). A comparison of techniques for nondestructive composition measurements in CdZnTe substrates. *JEM, 24*(5), 697–705. https://doi.org/10.1007/BF02657981

30. Bolotnikov, A. E., et al. (2007, October). Effects of Te inclusions on the performance of CdZnTe radiation detectors. In *2007 IEEE nuclear science symposium conference record* (Vol. 3, pp. 1788–1797). https://doi.org/10.1109/NSSMIC.2007.4436507
31. Fiederle, M., Fauler, A., & Zwerger, A. (2007, August). Crystal growth and characterization of detector grade (Cd, Zn)Te crystals. *IEEE Transactions on Nuclear Science, 54*(4), 769–772. https://doi.org/10.1109/TNS.2007.902352
32. Zou, J., Fauler, A., Senchenkov, A. S., Kolesnikov, N. N., & Fiederle, M. (2021, June). Analysis of Te inclusion striations in (Cd, Zn)Te crystals grown by traveling heater method. *Crystals, 11*(6), 6. https://doi.org/10.3390/cryst11060649
33. Henager, C. H., Alvine, K. J., Bliss, M., Riley, B. J., & Stave, J. A. (2015, November). The influence of constitutional supercooling on the distribution of Te-particles in melt-grown CZT. *Journal of Electronic Materials, 44*(11), 4604–4621. https://doi.org/10.1007/s11664-015-3995-y
34. McCoy, J. J., Kakkireni, S., Gélinas, G., Garaffa, J. F., Swain, S. K., & Lynn, K. G. (2020, April). Effects of excess Te on flux inclusion formation in the growth of cadmium zinc telluride when forced melt convection is applied. *Journal of Crystal Growth, 535*, 125542. https://doi.org/10.1016/j.jcrysgro.2020.125542
35. Tilli, M. (2015). Chapter 4 – Silicon wafers: Preparation and properties. In M. Tilli, T. Motooka, V.-M. Airaksinen, S. Franssila, M. Paulasto-Kröckel, & V. Lindroos (Eds.), *Handbook of silicon based MEMS materials and technologies* (2nd ed., pp. 86–103). William Andrew Publishing. https://doi.org/10.1016/B978-0-323-29965-7.00004-X
36. Camp, P. R. (Apr. 1954). Resistivity striations in germanium crystals. *Journal of Applied Physics, 25*(4), 459–463. https://doi.org/10.1063/1.1721662
37. Merrell, J. L., et al. (2016, May). Enabling on-axis InSb crystal growth for high-volume wafer production: characterizing and eliminating variation in electrical performance for IR focal plane array applications. In *Infrared technology and applications XLII* (Vol. 9819, p. 981915). https://doi.org/10.1117/12.2223956
38. Le Donne, A., Binetti, S., Folegatti, V., & Coletti, G. (2016, July). On the nature of striations in n-type silicon solar cells. *Applied Physics Letters, 109*(3), 033907. https://doi.org/10.1063/1.4959558
39. Chernov, A. A., & Temkin, D. E. (1977). Capture of inclusions in crystal growth. In current topics of materials science: Crystal growth and materials; current topics of materials science: Amsterdam. *The Netherlands, 2*, 3–77.
40. Zázvorka, J., et al. (2019, June). Inhomogeneous resistivity and its effect on CdZnTe-based radiation detectors operating at high radiation fluxes. *Journal of Physics D: Applied Physics, 52*(32), 325109. https://doi.org/10.1088/1361-6463/ab23e3
41. Zázvorka, J., Franc, J., Hlídek, P., & Grill, R. (2013, November). Photoluminescence spectroscopy of semi-insulating CdZnTe and its correlation to resistivity and photoconductivity. *Journal of Luminescence, 143*, 382–387. https://doi.org/10.1016/j.jlumin.2013.05.042
42. Kim, K. H., Na, Y. H., Park, Y. J., Jung, T. R., Kim, S. U., & Hong, J. K. (2004, December). Characterization of high-resistivity poly-CdZnTe thick films grown by thermal evaporation method. *IEEE Transactions on Nuclear Science, 51*(6), 3094–3097. https://doi.org/10.1109/TNS.2004.839084
43. Benz, K. W., & Dold, P. (2002, April). Crystal growth under microgravity: Present results and future prospects towards the International Space Station. *Journal of Crystal Growth, 237–239*, 1638–1645. https://doi.org/10.1016/S0022-0248(01)02358-2

Solution Growth of CdZnTe Crystals for X-Ray Detector

Song Zhang, Bangzhao Hong, Lili Zheng, Hui Zhang, Cheng Wang, and Bo Zhao

1 Introduction

The development of sensitive and portable X-ray and gamma radiation detectors needs progress in production of large, high-quality crystalline cadmium zinc telluride (CdZnTe). The modified vertical Bridgman method has been widely used as a melt growth method in preparation of CdZnTe crystals [1, 2]. However, this melt growth method suffers from high growth temperature and segregation of solute, which cause common problems in crystals, such as extended structural defects and inhomogeneity. Solution growth method is proposed as a low-temperature growth method, which is attracting more and more attentions during the studies of detector-grade CdZnTe crystals [3–5]. Commonly, the CdZnTe crystals are prepared from a Te-rich solution in solution growth method which has the potential to improve the problems during melt growth method [6]. There are two main solution growth methods widely used, i.e., the temperature gradient solution growth method and the traveling heater method.

The temperature gradient solution growth (TGSG) method as shown in Fig. 1a is proposed by combing solution growth method and vertical Bridgman method during which the structure of furnace is like a VB furnace, while the feeding material is a Te-rich solution so that the temperature of growth interface can be reduced. The interface temperature is determined by constitution of the remaining solution based on the phase diagram. During TGSG process, all the feeding material is melted and mixed before pulling of the ampoule. The interface temperature is reduced with

S. Zhang (✉) · B. Hong · L. Zheng · H. Zhang
Tsinghua University, Beijing, China
e-mail: zsthu@tsinghua.edu.cn

C. Wang · B. Zhao
Ruiyan Technology Co. Ltd, Chengdu, China

© The Author(s), under exclusive license to Springer Nature Switzerland AG 2023
L. Abbene, K. (Kris) Iniewski (eds.), *High-Z Materials for X-ray Detection*,
https://doi.org/10.1007/978-3-031-20955-0_4

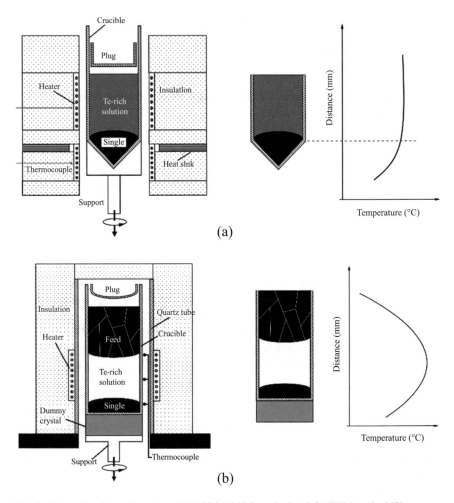

Fig. 1 Schematic of growth system, (**a**) TGSG/DMSG method and (**b**) THM method [7]

reduction of solute in the Te-rich solution, for which the temperature of heaters is also controlled to reduce so as to maintain the position of growth interface. The traveling heater method (THM) is quite similar to zone melting (see Fig. 1b), where the pre-synthesized polycrystalline feeding dissolves into Te-rich solution through diffusion and convective transport and finally deposits at the growth interface when the heater moves up (or more commonly, the ampoule moves downward). Compared with melt growth, the traveling heater method is a suitable solution growth method for growing crystals with high melting point or high vapor pressure and has its intrinsic advantages, such as low growth temperature, charge purification, and longitudinal homogeneity [6].

To date, the mass production of CdZnTe single crystals larger than 20 mm × 20 mm × 15 mm with low defects and uniform composition is still a big challenge and suffers from a low yield rate. To be noted, Te inclusion is still an important defect that degrades material properties during solution growth method, which is caused by constitutional supercooling and unstable growth interface, resulting in random trapping of Te-rich solution [8]. For growing high-quality grains, it is essential not only to create optimal crystal growth conditions but also to ensure the stability of thermal and flow fields during crystal growth. However, there has been little discussion on maintaining a stable thermal environment in CdZnTe crystal growth. Another important issue for solution growth is the preparation of raw materials including seed, Te-rich solution, and feeding material. This makes the growth process complicated and also increases the possibility of contamination. The same problem occurs for the as-grown crystals for both TGSG and THM methods.

This chapter presents an overview of our efforts in solution growth of large size $Cd_{0.9}Zn_{0.1}Te$ crystals for X-ray detectors. Firstly, a direct mixing solution growth method is proposed to prepare high-purity CdZnTe crystals from 7N raw material during one single step, and the accelerated crucible rotation technique (ACRT) is used to improve the crystal size and quality. Secondly, unseeded THM process is studied to optimize the THM process. In this part, the THM furnace is firstly designed based on numerical investigation of the relationships among the ampoule diameter, the height of Te-rich solution, and the height of heater. The thermal field stability during crystal growth is also improved by introducing a dummy crystal which has the same thermal conductivity with CdZnTe crystals. Controlling of growth interface by ampoule ration is also studied in order to get large crystals through unseeded THM growth. Finally, all the crystals are characterized by IR transmission images of Te inclusion, composition, and I–V curve of the as-made CdZnTe detectors. A comparison between the DMSG and THM method is conducted based on numerical simulation and characterization of as-grown CdZnTe crystals, and some future research work to improve the crystal quality is discussed at the end of this chapter.

2 Direct Mixed Solution Growth Method

2.1 Design of Direct Mixed Solution Growth Process

The direct mixed solution growth method is a one-step method based on solution growth method which means that the preparation of CdZnTe feeding material and the growth process is combined together during the synthesis of CdZnTe single crystal [9]. Since no seed is used, the material preparation and crystal growth are achieved by enlargement of random nucleation. The growth system used for DMSG method is shown in Fig. 1a, and a vertically increasing thermal profile like VB method is adopted. The furnace is a self-designed and homemade two-zone electric furnace, and two heaters are separated by insulation brick and a heat sink. Modular design is used in the design of furnace so that the thermal field can be easily modified by changing parts of the furnace. The upper heater is the main heater which controls the temperature profile of the melting region. The lower heater and the heat sink are designed to modify the post-growth profile which can also act as a post-growth annealing process. The ampoule is placed onto the ampoule support which can control the movement of ampoule with stepping motors. The temperature profile used in the experiment is shown in Fig. 1a, and the temperature at the growth interface is 890 °C which is tested using an external thermocouple during the experiments. The temperature gradient at the growth interface can be controlled by changing the set point of the upper and lower heaters and also the position of growth interface. For example, by increasing the temperature of the upper heater and lowering the ampoule, a higher temperature gradient in the solution can be obtained.

Carbon-coated quartz ampoule of 75 mm in diameter with a 120° cone-shaped bottom is used. The ampoule is filled with the Cd, Te, and Zn raw materials and In as dopant and then vacuum sealed before the growth process. The raw material of Cd, Te, and Zn are directly mixed to form the Te-rich solution. By controlling the reaction process of raw materials, the heat release rate is controlled, thus avoiding explosion of sealed ampoule. The reaction inside the ampoule can be indicated by a rapid temperature increase of thermocouple at wall of ampoule. As shown in Fig. 2, reaction occurs at 450 and 650 °C. Before growth, the ampoule is raised to the position with temperature of 890 °C. At the saturation point of the source material, CdZnTe is fully mixed in the Te-rich solution by flow in the ampoule which is derived from temperature gradient and rotation of ampoule. After the source material is fully dissolved, the ampoule is pulled downward at a predetermined speed. The initial growth temperature is 890 °C, and the temperature of growth interface is lowered with the decreasing of CdZnTe solute, which is controlled as a function of crystal growth rate and time.

2.2   Optimizations of Direct Mixed Solution Growth Process

Since the vertical temperature profile is monotonically increasing, which is a stable thermal structure for convection, the convection of solution is mainly caused by the lateral temperature difference across the ampoule. Figure 3a shows the numerical simulation of thermal and flow field in a typical DMSG process. It is seen that the main vortex is formed at the top of solution and there is slight flow near the growth interface. The solution near the growth front is quite stable, for which it is easy to control the growth interface by changing the temperature field. However, the heat and mass transfer are also in a low level without solution flow, which leads to a low temperature gradient and low crystal growth rate. In order to enhance the heat and mass transfer near the growth interface, rotation of ampoule is used in the mixing stage before crystal growth.

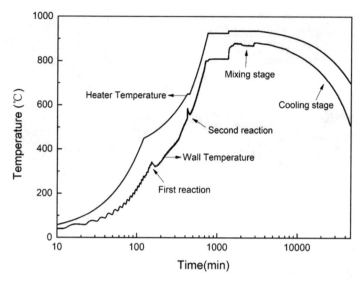

Fig. 2 Temperature evolution during DMSG process

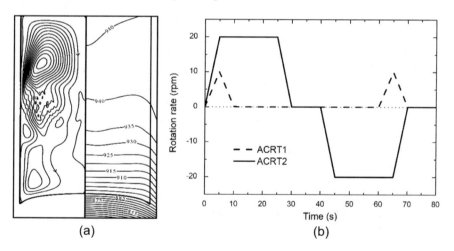

Fig. 3 (**a**) Numerical simulation of temperature and flow field in DMSG, (**b**) ACRT sequence used to enhance the heat and mass transfer in DMSG

Figure 4 shows four typical crystal growth experiments conducted in order to improve the crystal quality with DMSG method. Two different ACRT processes are introduced (see Fig. 3b), during which the Ekman flow will occur during acceleration and deceleration and cause mixing near the solid/fluid interfaces. A high rotation rate with a high frequency is necessary to destroy the stable density stratification and facilitate mixing, due to the stable thermal structure near the growth interface. Crystal sample 1 is grown at the temperature of 890 °C, and

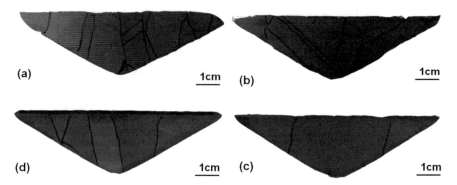

Fig. 4 Longitudinal cross section of as-grown crystals in DMSG experiments [9]. (**a**) Sample 1, (**b**) sample 2, (**c**) sample 3, (**d**) sample 4

the temperature gradient is measured relatively low at 20 K/cm. The growth rate is 3 mm/day for the first 2 days at the nucleation stage, after which the growth rate is kept at 5 mm/day. Longitudinal cross sections of the as-grown crystal (Fig. 4a) show that initial grains are formed at the tip of cone, but such grains disappear soon and new grains are formed at the wall of ampoule. This leads to formation of many small grains in sample 1. During the growth of sample 2 (see Fig. 4b), the growth interface temperature gradient is increased to 40 K/cm, while growth temperature and growth rate remain the same. However, the grain size is almost not improved. An ACRT pattern (dash line in Fig. 3b) is introduced during the growth of sample 3. From the cross sections of the as-grown sample 3, it is observed that initial grains are formed at the tip of cone and grow continuously further without too many new grains formed at the wall of ampoule (see Fig. 4c). Another bidirectional ACRT process is used in growth of crystal sample 4 which is shown with solid line in Fig. 3b. The longitude cross section of the as-grown crystal in Fig. 4d shows that initial grain is also formed at the top of the ampoule and grows continuously to the end of the growth process. However, there are still unexpected new grains formed during the enlarging process. The cross section of the crystal shows that the grain size is improved than sample 3 and the number of unexpected small grains is reduced obviously. From the cross section of samples, it is seen that the introduction of ACRT process can improve the crystal size during DMSG growth of CdZnTe crystal.

Increasing temperature gradient at the growth interface makes the growth process more stable when there is a fluctuation of thermal field. In sample 3, there is a periodic stir of the solution to break the balance built by buoyance flow. A more intense flow will increase the heat transfer in the solution. In sample 4, a bidirectional ACRT is used to increase the forced flow, and this will compress the supercooling layer at the growth interface by increasing of temperature gradient. By enhancing temperature gradient and mass transfer rate at growth interface with ACRT, large grains can be easily achieved in DMSG method. However, there is an important problem of MDSG method, i.e., the concentration nonuniformity of dopant elements. From Fig. 5 it is shown that the axial concentration of indium in

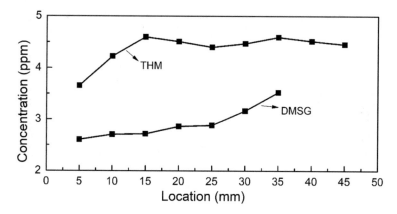

Fig. 5 Indium concentration along the growth direction during DMSG and unseeded THM method

recently prepared DMSG ingot remains uniform at the beginning of growth process and increases more and more rapidly towards the end of the ingot. This phenomenon is mainly caused by the segregation effect of indium. Another potential reason for the nonuniformity is the changing growth temperature during the growth process.

The resistivity at different locations in DMSG prepared ingot is characterized after annealing in a Te-Cd atmosphere (see Fig. 6). It is shown that the resistivity at the tip of the ingot is quite stable near 2.0×10^{10} Ω·cm under a bias of -600 to $+600$ V. However, the resistivity reaches 1.7×10^{12} Ω·cm under a bias of -600 to 0 V while 3.5×10^{10} Ω·cm under a bias of 0 to 600 V at the tail of the ingot. As shown in Fig. 5, the dopant concentration is different between two faces of the wafers prepared by DMSG method, and this is much more severe at the tail part of the ingot. Combining the resistivity analysis, the nonuniformity of dopant concentration is a potential reason for the asymmetric resistivity distribution, even though this is commonly caused by a poor ohmic contacts between the contact layer and CdZnTe samples.

Figure 7 shows Te inclusions in DMSG prepared crystals before (Fig. 7a–c) and after (Fig. 7d) annealing which is captured using IR transmission microscopy. There are three types of Te inclusion formed during DMSG growth process, i.e., 1) large amount of small Te inclusions with the size smaller than 3 μm, which is shown in Figs. 2 and 7a, c) small-size Te inclusions (smaller than 3 μm) along the line of dislocation and twin boundary as shown in Figs. 3 and 7b, c) occasionally formed large-size triangular or hexagon Te inclusions with the size between 15 and 20 μm. Since the size of most Te inclusion is smaller than 3 μm and there are huge number of this kind of inclusion, the density of Te inclusion is not counted. After an annealing process under Te-Cd atmosphere at 600 °C for 80 h, Te inclusions are improved as shown in Fig. 7d. After annealing, there are only inclusions with the size near 5 μm that can be clearly seen, and the large-size inclusions disappear. Also, there will be several remaining small-size Te inclusions which is located at the

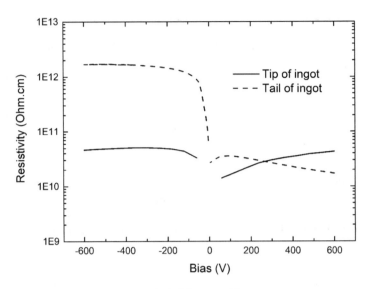

Fig. 6 Resistivity at different locations of DMSG prepared ingot

dislocation before annealing. The source of the remaining 5-μm-large inclusions is still unknown, and there is no evidence that can indicate the remaining inclusion is formed by reunion of small size inclusions or by shrinkage of large-size inclusion (the density of remaining inclusions is quite larger than large-size inclusions before annealing).

3   Traveling Heater Method

As discussed in DMSG growth process, a stable and high temperature gradient near the growth interface is essential to prepare large and high-quality grains. The growth conditions could be stable in THM method when balance between melting of feeding material and solidification at the growth interface is built, and thus the longitudinal homogeneity can be hugely improved. However, as shown in Fig. 8, a vortex with an upward flow near the ampoule wall and a downward flow along the center occupies the main area of molten zone due to the buoyancy effects, while diffusion works only in thin boundary layers. Such convection, arising from a parabolic temperature profile, plays a significant role in enhancing mixing, but, on the other hand, is considered as the main cause of concave growth interfaces [10, 11]. It is well-known that a slightly convex growth interface is essential for grain enlargement and preventing polycrystalline growth through post-formed grains on the ampoule wall. In order to obtain a convex interface, many researches start by weakening natural convection using a shorter molten zone, ampoule rotation, an additional magnetic field, and microgravity growth [12–15]. Ampoule rotation is

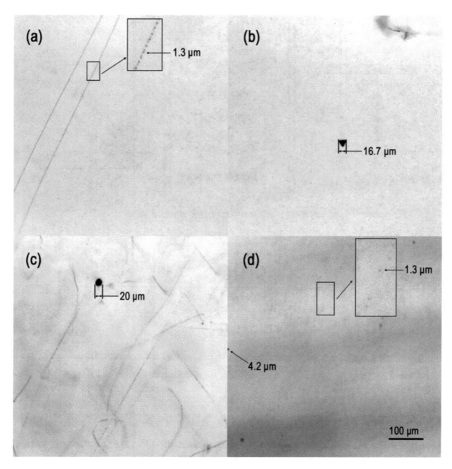

Fig. 7 IR transmission images of CdZnTe prepared in DMSG method, (**a–c**) as-grown crystals, (**d**) after annealing

a simple and convenient technology in experiments. Lan [16] first reported that a concave growth front due to buoyancy convection can be easily inverted into a convex one by applying a constant rotation of 3–5 RPM for THM growth of 1-inch GaAs. Dost [17] also found that a constant rotation of 5.0 RPM for a 1-inch CdTe crystal is effective in optimizing the interface. However, there have been no further experimental verifications.

Aiming for a stable and convex growth front with relatively large axial temperature gradient, fast unseeded THM growth of $Cd_{0.9}Zn_{0.1}Te$ crystals is conducted by introducing DMSG growth during feeding material preparation step. The procedures of fast unseeded THM growth are shown in Fig. 9. The schematic diagram of the THM growth system is shown in Fig. 1b. The furnace is a self-designed one-zone electric furnace with a precision of $\pm 0.1\ °C$. Before growth, polycrystalline

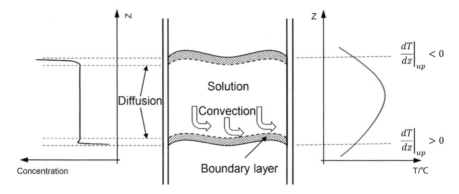

Fig. 8 Schematic of heat and mass transfer regime in the solution area

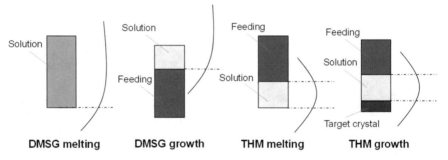

Fig. 9 Procedures of fast unseeded THM growth

CdZnTe feed is synthesized from 7 N metals using the DMSG method with a composition of Cd:Zn = 9:1. Then, the pre-synthesized CdZnTe polycrystalline with pure tellurium as solvent material and indium as dopant are loaded into a carbon-coated quartz ampoule and vacuum sealed at 5×10^{-4} Pa. The growth temperature is designed at 750 °C by controlling the amount of excess tellurium. The sealed ampoule and a dummy crystal are orderly placed on the support before growth. Three thermocouples are fixed at different locations on the ampoule wall to monitor the initial temperature evolution at nucleation stage.

Several CdZnTe ingots with different diameters are prepared through fast unseeded THM as listed in Table 1. A relatively low growth rate of 3 mm/day is used to control nucleation, which is subsequently increased to 5 mm/day for routine growth. In order to capture the growth interface shape, ingots are rapidly cooled to room temperature and then cut along the axial direction. Based on simulation results, the thermal structure of 2-in and 3-in THM furnace is redesigned to obtain a stable and high-temperature gradient at the growth interface. On the basis of stable thermal profile, the control strategy for interface shape is discussed by analysis of numerical simulation and experimental results.

Table 1 Growth process parameters used in fast unseeded THM experiments

Case#	Growth rate (mm/day)	Diameter (mm)	Length of grown crystal (mm)	Constant rotation rate (RPM)	Dummy crystal
CZT-1	3~5	27	27	/	NO
CZT-2	3~5	27	27	/	Yes
CZT-3	3~5	50	21	2.5	Yes
CZT-4	3~5	75	10	0	Yes
CZT-5	3~5	75	15	1.25	Yes

3.1   Thermal Optimization

3.1.1   Design of Thermal Structure

As shown in Fig. 1b, a one-zone furnace is used to build the parabolic temperature profile which is needed for the THM growth. The main conditions for stable growth of THM are a parabolic temperature profile with large axial temperature gradient, a slightly convex growth front, and a stable melt flow. Generally, these objectives are not mutually promoted, but contradictory. For instance, because of the parabolic temperature profile used in THM, the buoyancy-induced flow is stronger than DMSG method. To avoid the constitutional supercooling, introducing a large temperature gradient will inevitably increase the temperature difference across the molten zone, which further strengthens the intensity of flow. For a given growth system, the growth conditions are basically determined by the thermal structure of furnace, including the heater, insulation material of furnace, ampoule, and the support of ampoule, and also there are other ways to optimize the growth conditions, such as rotation of ampoule. Normally, homogeneous insulation material is used to build the furnace body, so the growth conditions are mainly determined by the ampoule diameter, the height of heater and the height of solution, and also coupling between the heater and support of ampoule.

The success of application of computational fluid dynamics to thermal design makes it capable to discuss the influence of different parameter combinations in a wider range. Numerical simulation is used to find the best relationship among the ampoule diameter, height of solution, and height of heater. Figure 10 shows the simulation results using our self-programmed software, MASTRAPP, for cases with the same growth interface temperature and different furnace structures, while the ampoule diameter is fixed. As shown in Fig. 10, the growth front, the axial temperature gradient, and the flow pattern are significantly changed under different combinations of heater and solution zone. In terms of interface shape, it is an effective way to weaken the melt flow by using a short solution zone, but there is a risk of interface touching. For long solution cases (see Fig. 10g–i), strong upward flow near the wall and downward flow near the centerline continually scour the upper polycrystalline and lower single crystal, respectively, which results in a convex dissolution interface and more concave growth interface. For cases $H/S < 0.57$ and $S/D > 1.05$, the flow pattern has changed greatly to form two vortexes

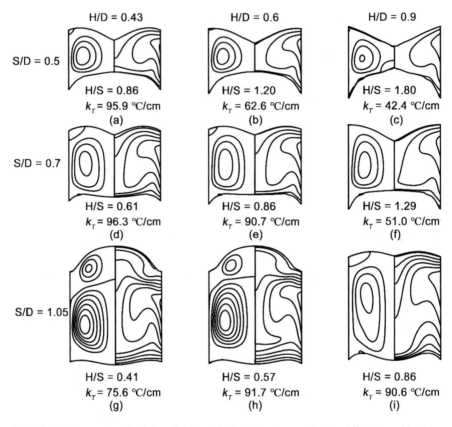

Fig. 10 The temperature field, flow field, and the interface shape under the different combinations of heater and solution zone. The left of each picture is streamline, and the right is isotherm. S, H, and D represent the height of solution zone, the height of heater, and ampoule diameter, respectively

in the solution zone. The upper vortex with counterclockwise direction occupies the region adjacent to the dissolution interface, which restricts the component transport from the dissolution interface to the growth interface. Another issue of using long solution zone is the potential smaller axial temperature gradient since the interface position is far away from the edge of heater.

For a certain height of solution zone, the height of heater mainly affects the temperature difference across the solution zone and the axial temperature gradient. Since the growth interface temperature is expected to remain unchanged, a higher heater temperature is needed to ensure the interface temperatures to reach the set value when using a short heater. This, from another viewpoint, increases the temperature difference across the solution zone and enhances the flow, resulting in more concave interface as shown in Fig. 10d–f. When the height of heater is larger than solution zone (see Fig. 10c–f), despite the temperature difference is decreased, the axial temperature gradient is also decreased since the solution zone is placed inside the heater in that cases.

It is concluded in Fig. 10 that S/D and H/S are two main parameters in determining the flow pattern and thermal conditions for THM system. For cases with H/S > 0.86, the interface temperature gradient tends to be much smaller since the solution zone is placed inside the heater. Cases with S/D > 1 are also not recommended due to the longer transport distance between the dissolution and growth interfaces, the complex flow pattern, the strong convection flow, and the concave growth interface. However, the main difficulty of using the short solution zone is to avoid the interface touching. Based on the simulation results in Fig. 10, the recommended structures for THM is 0.5 < S/D < 0.7 and 0.61 < H/S < 0.86, while the exact limits of each parameter are not studied in this chapter.

3.1.2 Temperature Profile Stability

Figure 11 shows the surface temperature of ampoule that is measured by three thermocouples located at the ampoule wall near solution zone for ingot CZT-1. The upper thermocouple and the lower thermocouple are located at the initial interface of molten zone, while the middle thermocouple is at the middle of molten zone (see Fig. 1b). Obviously, the ampoule wall temperature gradually increases with the pulling process of ampoule. For instance, at position of 115 mm, the temperature difference between the upper and lower thermocouples reaches 37 °C. This is because of the deposition of low-thermal-conductivity CdZnTe on the bottom of ampoule, which blocks the heat extraction from crucible and makes the thermal field increased. The increasing temperature is believed to enhance the convection intensity, even change the flow structure, which further affects the concentration distribution, interface shape, and growth temperature. Figure 12 shows the evolution of growth interface with different lengths (L) of single crystal deposited on ampoule bottom. Because of increasing wall temperature, the average growth temperature is also increased from 748.6 to 766.0 °C in simulation. Since the growth interface gradually moves away from heater and the increase of flow intensity, the growth interface shape also changes from convex to concave. The reduction of grain size and post-formed grains are considered to be related to the transition of interface shape from convex to concave.

The key factor of stabilizing the thermal field growth process is to maintain the stability of heat release near ampoule bottom at the early stage when low-thermal-conductivity CdZnTe crystals accumulate. As presented in Fig. 12b, such temperature rise will disappear after a certain length of single crystal deposited on ampoule bottom. Based on the analysis, a simple and easy-to-process support is designed [7] to stabilize the thermal field by choosing appropriate material support. Figure 11b is the temperature profile of the ampoule surface in crystal growth of ingot CZT-2. It is obvious that the temperature of ampoule wall does not change during the pulling process of growth. By introducing the dummy crystal as a support of ampoule, thermal decoupling between ampoule and support is achieved, i.e., a

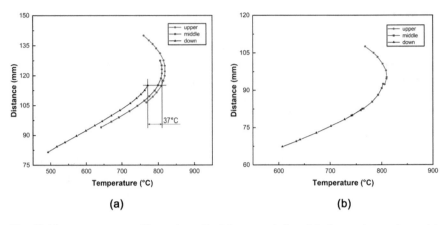

Fig. 11 The temperature profile on the wall of the ampoule in a 1-inch growth experiment (**a**) (CZT-1) and (**b**) (CZT-2) a dummy crystal [7]

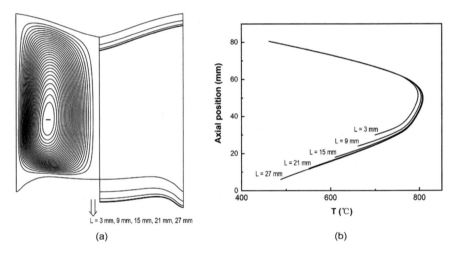

Fig. 12 (**a**) The left-hand side of the picture is streamlines at $L = 3$ mm, while the growth and dissolution interfaces under increasing thermal field are on the right-hand side. (**b**) The simulated wall temperature profiles [7]

stable thermal condition is built at the beginning of the nucleation stage, during which the change of temperature condition caused by changes in thermal structure is the main problem.

3.2 Interface Control by Rotation of Ampoule

For a growth system without rotation, buoyancy is the only driving force of melt flow. As discussed before, an upward flow near the wall and a downward flow near the centerline continuously scour the feeding and the already grown crystals, respectively, which results in a convex dissolution interface and a concave growth interface. Such interfaces are clearly observed in the 3-inch experiment (CZT-4), as shown in Fig. 14b. Figure 13 shows the effects of constant rotation on flow and thermal fields in a 2-inch THM system. Fundamental changes occur in the flow structure after applying constant rotation, even at 1.0 RPM (see Fig. 13b). The upward flow arises near the wall of ampoule because of the temperature difference and turns inward after reaching the feed crystal. The inward melt flow is weakened after flowing through the dissolution interface in the presence of centrifugal force. After turning downward, the downward flow is further weakened because of the positive temperature gradient at the growth front and begins to flow outward under the action of centrifugal force. As a consequence, natural convection, which initially occupies the core area of the molten zone, is squeezed to the upper left part (see Fig. 13c), and a counter-rotating vortex arises in the center of the growth interface, which makes the interface convex. However, Fig. 13d shows that a further increase in rotation rate has limited positive effect on the growth interface shape or even makes it worse. The primary large vortex breaks into many small vortices, which may weaken the transport of components between the external region near the wall and the internal region near the center. Figure 14a shows the convex growth interface obtained in a 2-inch THM growth system (CZT-3) at a constant rotation rate of 2.5 RPM, while the grains are enlarging. At the same time, the simulated growth and dissolution interfaces shape are quite similar to the experimental results.

Such an interface transition from concave to convex is the consequence of the confrontation between thermal convection and forced convection, which has also been observed in Czochralski growth of garnets [18, 19]. With an increase in rotation rate, the intensity of thermal convection does not decrease, while the Grashof number fluctuates around 8.5×10^5. On the other hand, the ratio of the Grashof number to the square of the Reynolds number (Gr/Re^2) decreases from 0.79 at 1.0 RPM to 0.13 at 2.5 RPM and 0.06 at 3.5 RPM. This value shows that it does not require a higher rotation rate to completely suppress thermal convection ($Gr/Re^2 <$ 0.1) with the purpose of controlling the growth interface shape.

A 3-inch ingot (CZT-5) is grown by THM to repeat these results in larger crystals. Since the Gr/Re^2 is proportional to $1/w^2R$, a smaller rotation rate than 2.5 RPM is assumed to be valid in a 3-inch THM growth system. From Fig. 14b, c, it is obvious that the growth interface changes from concave to convex, even though only 1.25 RPM is used. In contrast to the elliptical convex growth interface in a 2-inch ingot, there is an obvious uplift in the center of growth interface, while a depression is formed on the uplift in the simulation (see Fig. 14c). The formation of uplift shows that the constant rotation of 1.25 RPM is still too large for a 3-inch THM growth system. Apart from making growth interface convex, on the other

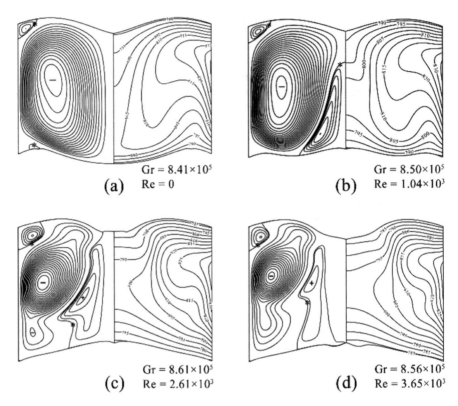

Fig. 13 The thermal and flow fields of a 2-inch crystal with different rotation rates: (**a**) 0.0 RPM, (**b**) 1.0 RPM, (**c**) 2.5 RPM, (**d**) 3.5 RPM. The streamlines are spaced at 1.2×10^{-6} for positive values, 1.2×10^{-7} for negative values, and the asterisk represents the zero streamline [7]

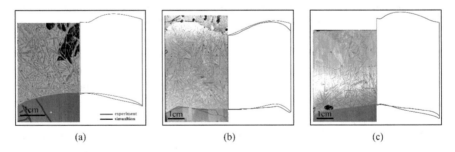

Fig. 14 Experimental and simulated interfaces of crystals: (**a**) 2-inch with 2.5 RPM, (**b**) 3-inch without rotation, (**c**) 3-inch with 1.25 RPM (end of growth) [7]

hand, a counter-rotating vortex leads to the accumulation of excess Te in the center. From the CdTe binary phase diagram, the liquidus temperature of crystals decreases with an increase in the Te amount, which results in a depression at the center of the growth interface. However, uncontrollable factors in the experiments, such as

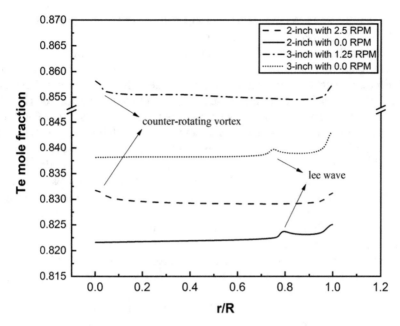

Fig. 15 The concentration of Te at the growth interface with different diameters and rotation rates [7]

geometric deviation, fluctuation of furnace temperature, eccentricity of rotation, and mechanical shaking, destroy the symmetry of two-dimensional model and, therefore, avoid the accumulation of Te to a certain extent.

Another concern in crystal growth is the uniformity of concentration in both axial and radial directions. Due to strong natural convection in THM, the radial distribution of Te is relatively uniform over most of the growth interface, while a counter-rotating vortex adjacent to the growth interface leads to the accumulation of Te in the local region, as shown in Fig. 15. Without rotation, the formation of lee wave [14] causes Te accumulation around 4/5R in 2-inch and 3-inch crystals. After introducing rotation, the original lee-wave structure disappears, and a counter-rotating vortex arises in the center, which moves Te-rich area to the center.

3.3   Numerical Study of ACRT in the THM Growth

The accelerated crucible rotation technique is a widely used method for the optimization of crystal growth. In traveling heater method, the interactions between the original natural convection and forced convection by ACRT make the flow behavior much more complicated. Since a small constant rotation of 2.5 RPM is sufficient to change the flow structure as proved above, we are interested in the influence of small-rotation ACRT. For comparison purposes, the maximum rotation rate is set as 2.5 RPM. As shown in Fig. 16b, relatively stable growth and dissolution interfaces are obtained in simulation. The growth interface is more convex toward the solution than the state without rotation but is not as convex as that under constant rotation at the same rotation rate. Although the position of the interfaces still fluctuates slightly over time in simulation due to the periodic interfacial heat fluxes, it's almost imperceptible, which is quite different from that of Lan [16] where a visible change in growth interface shape appeared in 25 s. From Fig. 16b, it is seen that there is no obvious Ekman flow near the solid/liquid interfaces. The formation of the counter-rotating vortex in the center of the growth interface at T/2 and T is more like the phenomenon occurring in the constant rotation case. Based on the criterion given in [20], the rotation rate for a stable Ekman flow in the situation of Fig. 16 is between 0.37 and 4.6 RPM, while the lower and upper limits of Re are 40 and 500, respectively. Possible reasons are the suppressing effect by natural convection and a long period of 50 sec used, which causes weak Ekman flows. Figure 16c shows a large transient growth rate that is two orders of magnitude larger than the pulling rate, which is also reflected in the evolution of temperature gradient (see Fig. 16d). Without the counter-rotating vortex, the convection carries heat directly to the growth interface, resulting in a large temperature gradient and a negative transient growth rate at the growth interface. When the vortex appears, the direction of natural convection is also changed, during which the heat fluxes through the center of the growth interface are greatly reduced. Consequently, a small temperature gradient appears in the center. The transient growth rate much greater than the pulling rate indicates that the rapid growth and remelting occur periodically during ACRT process.

A larger rotation rate is needed to enhance the intensity of Ekman flow and avoid the formation of the centrifugal-force-induced vortex. When the Gr number is about 8×10^5 in the initial state, if the forced convection was the dominant factor, a rotation rate greater than 26 RPM is needed so that $Gr/Re^2 < 0.1$. Figure 17 shows the flow field, thermal field, and interface shape during the ACRT with a maximum rotation rate of 15 RPM. The coupling of the natural convection and Ekman flows makes the flow more complicated. During acceleration, a counter-rotating Ekman flow is generated at the dissolution interface, which occupies the upper-left corner of the solution zone. At the same time, the Ekman flow formed at the growth interface is merged with the natural convection because of the same flow direction as shown by the distortion of stream function at the lower-left corner of the solution zone in Fig. 17b. During deceleration, the flows near the growth interface are greatly suppressed. This is because the direction of inward Ekman flow by deceleration is opposite to the flow after completing acceleration. It is worth mentioning that even though the Ekman flow is enhanced, the core region of the solution zone is still less affected. For the region adjacent to the growth interface, such alternating Ekman flows only work in the periphery of the growth interface, which causes a fluctuating temperature gradient in the local region (see Fig. 17d).

As shown in Fig. 17b, there is no natural convective flow directly scouring the growth interface at these four typical moments, which results in a convex growth interface. The transient growth rate is better than that in small-rotation ACRT as

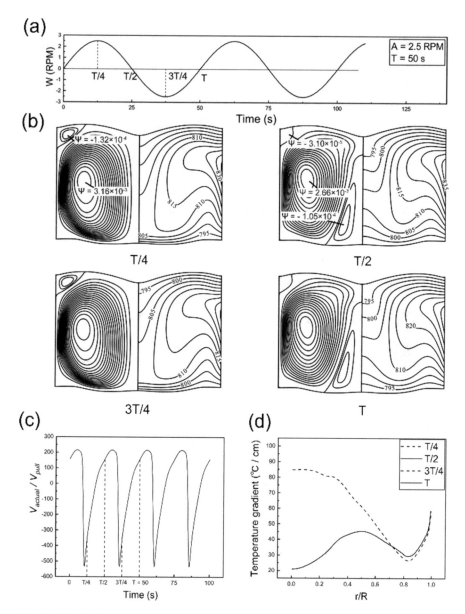

Fig. 16 The simulation results of 2.5 RPM ACRT with a period of 50s: (**a**) the ACRT regime; (**b**) the thermal and flow fields in one cycle. (**c**) Variation of the ratio of transient growth rate at $r/R = 0$ on the growth interface to pulling rate with time while V_{pull} is the pulling rate. (**d**) Variation of temperature gradient along the growth interface with time while the curves at T/4 and 3T/4 coincide with each other, as well as at T/2 and T

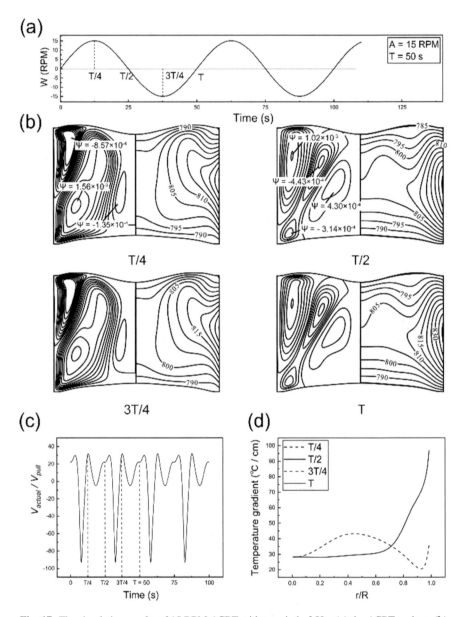

Fig. 17 The simulation results of 15 RPM ACRT with a period of 50s: (**a**) the ACRT regime; (**b**) the thermal and flow fields in one cycle. (**c**) Variation of the ratio of transient growth rate at $r/R = 0$ on the growth interface to pulling rate with time. (**d**) Variation of temperature gradient along the growth interface with time while the curves at T/4 and 3T/4 coincide with each other, as well as at T/2 and T

shown in Fig. 17c, although it is still an order of magnitude larger than the pulling rate. A large negative transient growth rate between 0 and T/4 indicates that during acceleration, there is a short period when the natural convection directly flows through the center of the growth interface, causing a fast remelting rate of the growth interface. After the formation of Ekman flow, the core region of the solution zone is replaced by a counter-rotating vortex as shown in Fig. 17b, which turns the transient growth rate into a positive value. For the temperature gradient (see Fig. 17d), the absolute value is small, around ~30 °C/cm, but it is more stable in the center than that in Fig. 16.

A maximum rotation rate is set as 40 RPM to simulate the influence of large-rotation ACRT as shown in Fig. 18. The solution zone is completely dominated by Ekman flows in this situation. Generally, the flow structure is similar to Fig. 17 during acceleration except for a separation between natural convection and the outward Ekman flow at the growth interface. During deceleration, several transient Couette flows are formed near the ampoule wall as shown in Fig. 18b. These time-dependent flows are believed to disrupt the existing concentration boundary layer and increase the mixing in the VB growth system [21, 22]. From the view of numerical simulation, the formation of such unstable flow increases the difficulty of convergence and destroys the periodicity of flow as shown by small differences in the flow fields and temperature gradient at T/4 and 3T/4 or T/2 and T. With the further increase of rotation rate, the growth interface is still convex, and the areas affected by Ekman flow are also extended. It is seen that the area of about 0.5–1.0 R of the growth interface is dominated by the Ekman flows with opposite flow directions during acceleration/deceleration. It is also possible to further extend this area by using a shorter acceleration/deceleration time and a larger rotation rate, of course. However, the rapid change of the flow direction due to a strong ACRT regime must be accompanied by a more fluctuating temperature gradient (see Fig. 18d), resulting in a large transient growth rate in the affected area. Figure 18c shows that the transient growth rate at the center of the growth interface increases again. That may be the potential shortcoming of using a large-rotation ACRT, although the effects of such a large transient growth rate still need further studies.

3.4 Characterization of THM Samples

As a comparison, the ingot prepared by THM (CZT-3 in Table 1) with a largest single-crystal volume of 3.26 cm^3 is characterized by concentration of indium, resistivity, and Te inclusion. The distribution of indium concentration for ingot CZT-3 is shown in Fig. 5. Different from DMSG, the concentration of indium increases at the early stage of the growth (the tip of ingot) and then keeps stable at 4.5 ppm when the stable status is built between melting and solidification. This is a great advantage of THM compared with DMSG method in controlling the uniformity of dopant. Since indium is dopant twice in both the feeding preparation process and the THM growth process. The concentration of dopant is higher than that of DMSG

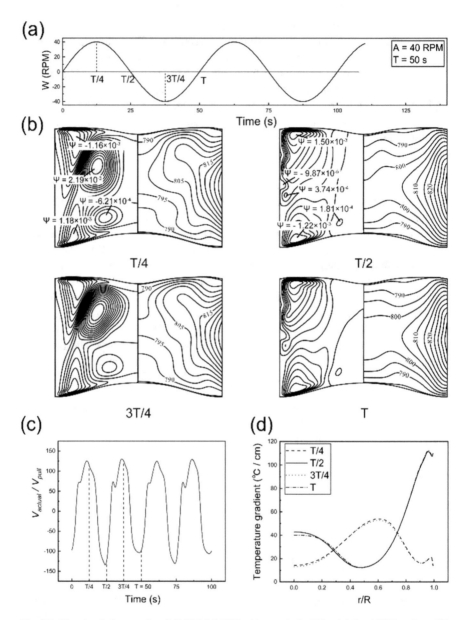

Fig. 18 The simulation results of 40 RPM ACRT with a period of 50s: (**a**) the ACRT regime; (**b**) the thermal and flow fields in one cycle. (**c**) Variation of the ratio of transient growth rate at $r/R = 0$ on the growth interface to pulling rate with time. (**d**) Variation of temperature gradient along the growth interface with time

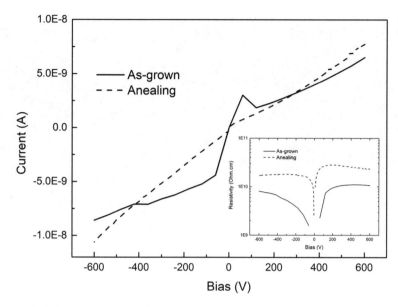

Fig. 19 The I–V curve of THM made crystals before and after annealing

process. The reason why there is an increase of dopant concentration is not fully understood since the concentration in the solution is quite high at the beginning of fast unseeded THM process.

In order to evaluate the THM-grown crystals, I–V curve of detector made by wafer at the stable part is measured and shown in Fig. 19. It is seen that the resistivity of the crystal is symmetrical due to the uniform of dopant concentration. However, the I–V curve is not smooth which indicates the resistivity is not stable under different bias. The resistivity of the as-grown crystal is 8×10^9 $\Omega \cdot$cm under a bias of -600 V and dropped to 1.6×10^9 $\Omega \cdot$cm quickly. Under the bias of $+60$ to $+600$ V, the resistivity of the as-grown crystal keeps at 1.0×10^{10} $\Omega \cdot$cm. The wafer is annealed under the same conditions as the case of DMSG in Fig. 6, i.e., an annealing process under a Te-Cd atmosphere at 600 °C for 80 h. From the I–V curve, it can be seen that the quality of the sample is improved on both the stability and the absolute value of the resistivity. The measured resistivity varies from 1.7×10^{10} $\Omega \cdot$cm to 2.7×10^{10} $\Omega \cdot$cm. By comparing the difference between resistivity of samples prepared by two methods, it can be seen that the resistivity is more stable under different bias in THM-grown crystals, but the absolute resistivity value of DMSG method is much higher than THM method. This phenomenon is thought to be caused by the difference in dopant concentration of the samples. Since the indium concentration of the THM sample is higher than the DMSG sample, the concentration of THM should be reduced. However, the optimization of doping process has not been studied. How to adjust the amount of indium in two doping processes is to be studied in the future fast unseeded THM process.

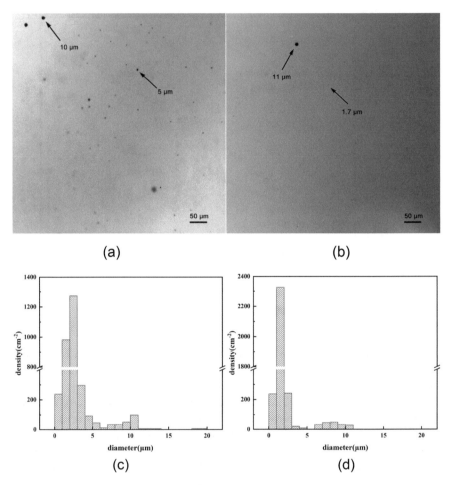

Fig. 20 Te inclusions in the THM samples (**a**), (**c**) before and (**b**), (**d**) after annealing

Additionally, Te inclusions in the THM samples before and after annealing are characterized with IR transmission images which are shown in Fig. 20. Figure 20a shows a typical image of Te inclusions of the as-grown crystals which contains Te inclusions with the size of 10 and 3–5 μm. The statistical data of Te inclusion counts in Fig. 20c shows that the inclusions are mainly distributed near 3 and 10 μm and there are also inclusions with large size of 20 μm. Figure 20b shows the results of Te inclusions after annealing. It is shown that there are still inclusions with the size near 10 μm, but most inclusions have quite small size of about 2 μm. The statistical data in Fig. 20d also show that the inclusions with the size near 5 and 10 μm are reduced and no inclusions larger than 15 μm are captured. The density of inclusions with the size of 2 μm is dramatically increased, which is also shown in Fig. 20b. However as mentioned by Chen [6], inclusions smaller than 3 μm with a concentration of $<10^6$ cm^{-3} can be tolerated in CdZnTe detectors with thickness up to 15 mm. Generally, annealing is an effective way to reduce the inclusions and improve the performance of THM-grown crystals.

4 Conclusion

Solution growth is an effective method for preparation of detector grade Cd$_{0.9}$Zn$_{0.1}$Te crystals. By controlling the reaction rate during synthesis process, the DMSG method is achieved which can prepare high-purity CdZnTe crystals from 7N raw material during one single step. Like the VB method, a stable temperature profile makes the convection quite weak in the melt which leads to easy control of the growth interface by changing the temperature field. However, the heat and mass transfer are also at quite low level without solution flow, which results in a low-temperature gradient and low crystal growth rate. Thus, ACRT with high rotation rate is used to get large size and high-quality CdZnTe crystals. The resistivity of annealed crystals reached higher than 10^{10} Ω·cm and even reached 10^{12} Ω·cm; however, the segregation effect leads to a nonuniform dopant concentration. And this restricts the application of DMSG in mass production of CdZnTe crystals, but this method can be used in preparing high-quality seeds and feedings for THM method. Based on DMSG method, fast unseeded THM process is conducted to improve the THM-growth system and optimize the process. The THM furnace is designed based on numerical investigation of the relationships among the ampoule diameter, the height of Te-rich solution, and the height of heater. The thermal field stability during crystal growth is also improved by introducing a dummy crystal which has the same thermal conductivity with CdZnTe crystals. Controlling of growth interface by ampoule rotation is also introduced in order to get large size crystals. The effects of different ACRT regimes during the THM growth are also studied providing a better understanding of interactions between the original natural convection and forced convection by ACRT, which is an alternative way to control the THM growth of CdZnTe. The resistivity of THM-prepared ingot can also reaches higher than 10^{10} Ω·cm after annealing under a Te-Cd atmosphere, which is more stable than that of DMSG method. Also, the annealing process can dramatically reduce the large size Te inclusions of the ingot and is an effective way to improve the quality of the crystal. However, there are still problems to be solved to improve the crystal quality:

1. Even though the resistivity is improved after annealing, there are still inclusions with the size near 10 μm left. How to furtherly reduce the Te inclusions through annealing has not been studied.
2. Based on the improved growth system, high-quality seeds can be used in the THM method to eliminate the random nucleation and achieve large-size crystal growth.

3. The concentration of indium dopant plays a determining role in the performance of detectors. Since the doping is conducted twice during THM process and a higher indium concentration is probably responsible for the low resistivity of THM than DMSG. The control strategy of indium concentration is not fully understood and needs a further investigation in the future work.

References

1. Li, L., et al. (2003). Studies of Cd-vacancies, indium dopant and impurities in CdZnTe crystals (Zn = 10%). In *2003 IEEE nuclear science symposium. Conference record.*
2. Xu, Y., et al. (2009). Characterization of CdZnTe crystals grown using a seeded modified vertical Bridgman method. *IEEE Transactions on Nuclear Science, 56*(5), 2808–2813.
3. Shiraki, H., et al. (2009). THM growth and characterization of 100 mm diameter CdTe single crystals. *IEEE Transactions on Nuclear Science, 56*(4), 1717–1723.
4. MacKenzie, J., et al. (2010). Recent advances in THM CZT for nuclear radiation detection. *Nuclear Radiation Detection Materials-2009, 1164*, 155.
5. Yin, L. Y., et al. (2017). The effects of ACRT on the growth of ZnTe crystal by the temperature gradient solution growth technique. *Crystals, 7*(3), 1–12.
6. Chen, H., et al. (2007). Characterization of traveling heater method (THM) grown Cd(0.9)Zn(0.1)Te crystals. *IEEE Transactions on Nuclear Science, 54*(4), 811–816.
7. Hong, B., et al. (2020). Studies on thermal and interface optimization for CdZnTe crystals by unseeded Traveling Heater Method. *Journal of Crystal Growth, 546*, 125776.
8. Roy, U. N. (2010). Macro- and microscopic growth interface study of CdZnTe ingots by THM technique, *7805*(1), 780502-780502-8.
9. Zhang, S., et al. (2018). Controlling Te inclusion during direct mixed solution growth of large size CdZnTe crystal. Proceedings of SPIE - The International Society for Optical Engineering 10762, 107620V
10. Roy, U. N., Burger, A., & James, R. B. (2013). Growth of CdZnTe crystals by the traveling heater method. *Journal of Crystal Growth, 379*(10), 57–62.
11. Liu, Y., et al. (2003). A three-dimensional numerical simulation model for the growth of CdTe single crystals by the travelling heater method under magnetic field. *Journal of Crystal Growth, 254*(3–4), 285–297.
12. Wang, Y., et al. (2005). Growth interface of CdZnTe grown from Te solution with THM technique under static magnetic field. *Journal of Crystal Growth, 284*(3–4), 406–411.
13. Roy, U. N. (2010). *Macro- and microscopic growth interface study of CdZnTe ingots by THM technique.*
14. Peterson, J. H., Fiederle, M., & Derby, J. J. (2016). Analysis of the traveling heater method for the growth of cadmium telluride. *Journal of Crystal Growth, 454*, 45–58.
15. Zhou, B. R., et al. (2018). Modification of growth interface of CdZnTe crystals in THM process by ACRT. *Journal of Crystal Growth, 483*, 281–284.
16. Lan, C. W., & Chian, J. H. (1999). Effects of ampoule rotation on vertical zone-melting crystal growth: Steady rotation versus accelerated crucible rotation technique (ACRT). *Journal of Crystal Growth, 203*(1), 286–296.
17. Dost, S., & Liu, Y. C. (2007). Controlling the growth interface shape in the growth of CdTe single crystals by the traveling heater method. *Comptes Rendus Mecanique, 335*(5–6), 323–329.
18. Zydzik, G. (1975). Interface transitions in Czochralski growth of garnets. *Materials Research Bulletin, 10*(7), 701–708.
19. Carruthers, J. R. (1976). Flow transitions and interface shapes in Czochralski growth of oxide crystals. *Journal of Crystal Growth, 36*(2), 212–214.

20. Brice, J. C., et al. (1986). ACRT - a review of models. *Progress in Crystal Growth and Characterization of Materials, 13*(3), 197–229.

21. Capper, P., Gosney, J. J. G., & Jones, C. L. (1984). Application of the accelerated crucible rotation technique to the Bridgman growth of CdxHg1-xTe: Simulations and crystal-growth. *Journal of Crystal Growth, 70*(1–2), 356–364.

22. Yeckel, A., & Derby, J. J. (2000). Effect of accelerated crucible rotation on melt composition in high-pressure vertical Bridgman growth of cadmium zinc telluride. *Journal of Crystal Growth, 209*(4), 734–750.

Three-Dimensional Mapping of Carrier Lifetime and Mobility

Selina R. H. Owe, Irfan Kuvvetli, and Carl Budtz-Jørgensen

1 Introduction

As CdZnTe (CZT) quality improved in the 1990s, the material became more favorable for the use of high-resolution room temperature detectors [8]. A great advantage of the heavy element semiconductor compound is the higher quantum efficiency achieved due to the larger atomic number. Furthermore, the large band gap energy allows for room temperature operation in contrast to germanium (Ge) detectors, which requires cryogenic cooling. These features are not only useful in industrial and medical imaging systems but are also essential features for detectors used in high-energy particle- and astrophysics [33]. CZT detectors are already implemented at several Space telescopes such as SWIFT [6], NuSTAR [26], and ASIM [4], and the material shows great promise for future Space missions. However, disadvantages are also associated with the material: ineffective charge collection due to charge trapping, difficulty of producing large-area defect-free single crystals, and material inhomogeneity. Furthermore, the hole movement in CZT is poor compared to electron movement, which impacts the spectral performance.

Different detector types exist for the CZT detector, mainly distinguished by the electrode geometry of the detector (e.g., planar, pixelated, strip). The pulse shape formation in the semiconductor detector will be affected by this electrode geometry but will also strongly depend on the material properties described by the mobility (μ) and lifetime (τ) of the charge carriers. The electrode geometry design chosen for the detector will affect the pulse shape formation, as described by the Shockley–Ramo theorem, due to the weighting potential distribution resulting from the given geometry [11, 25, 30]. The material properties μ and τ, however, are not a result

S. R. H. Owe (✉) · I. Kuvvetli · C. Budtz-Jørgensen
DTU Space, Technical University of Denmark, Lyngby, Denmark
e-mail: shoowe@space.dtu.dk

© The Author(s), under exclusive license to Springer Nature Switzerland AG 2023
L. Abbene, K. (Kris) Iniewski (eds.), *High-Z Materials for X-ray Detection*,
https://doi.org/10.1007/978-3-031-20955-0_5

of the given design of the detector, but of the given detector material, and the quality hereof. To truly describe the detector and its performance, it is important to understand the movement of charge carriers inside the material, from the point of creation to the point of collection. This includes understanding not only the signal formation due to the geometry of the electrodes but also the material properties μ and τ and how they affect the signal formation. In the following sections, we will give a review of some commonly used techniques to evaluate μ and τ of CZT material and how some of these techniques can be used to map the material properties of a CZT detector in 3D.

2 Review of $\mu\tau$ Extraction Methods

A relationship between the detector performance and the material parameters can be described through the mobility, μ, and lifetime, τ, of the generated charge carriers. μ has the unit $\left[\mathrm{cm}^2/\mathrm{V} \cdot \mathrm{s}\right]$ and describes how the motion of a charge is influenced by an applied electric field [32]. The larger the value of μ, the better mobility of the charge carriers in the bulk of the material. μ will decrease as the impurity concentration in the material increases [27]. τ has the unit [s] and is a measure of the trapping rate of the charge carriers, drifting from their origin toward the collecting electrodes [2]. When comparing the quality of CZT material, the product of the two parameters is often used as a measure referred to as the $\mu\tau$-product. Two carriers are generated within CZT as radiation interacts with the material, namely electrons and holes. These have different mobilities and lifetimes in the material, and therefore these must be evaluated separately. The $\mu\tau$-products of electrons and holes are referred to as $\mu_e\tau_e$ and $\mu_h\tau_h$, respectively.

Different characterization methods exist to understand the bulk properties of a semiconductor material, independent of the surface properties. Many of these transient techniques involve analyzing pulse shapes (current or charge signals) produced by irradiating the detector material with a laser pulse, α particles, or an X-ray source [27]. In the following sections, we will present some commonly used techniques for characterizing wide band gap semiconductor detector materials.

2.1 Fitting the Hecht Equation

A conventional method for determining the $\mu\tau$-product of a single carrier is based on the Hecht equation [16]. In this method, the charge collection efficiency (CCE) of the detector is utilized.

In 1932, K. Hecht defined a model describing the amount of charge which can be collected at the anode surface of a crystal illuminated by radiation [16]. The equation describes the relation between the collected charge and the initially generated charge by photon interaction, as a function of carrier drift length. This is also what can be described as the Charge Collection Efficiency (CCE). The Hecht equation exists as a one-carrier equation only describing the electrons/holes alone and a two-carrier equation describing both electrons and holes. The two-carrier Hecht equation is given by

$$Q = eN_0 \left\{ \frac{\lambda_e}{D} \left[1 - \exp\left(\frac{x_i - D}{\lambda_e} \right) \right] + \frac{\lambda_h}{D} \left[1 - \exp\left(\frac{-x_i}{\lambda_h} \right) \right] \right\},$$ (1)

where N_0 is the initial number of electron–hole pairs generated by the incident radiation, e is the electronic charge, x_i is the radiation interaction location measured from the cathode, D is the detector thickness, λ is the mean free path, and the subscripts e and h indicate electrons and holes, respectively [17]. With an applied electric field, E, the mean free path, λ, of the carriers for a given material is given by

$$\lambda = \mu \tau E.$$ (2)

The mean free path is a measure of an average distance a carrier can travel inside the detector before it is captured. Therefore, this measure should ideally be greater than the detector thickness itself for optimal energy resolution [27].

The Hecht equation can then be rewritten in terms of the $\mu\tau$-product as

$$Q = eN_0 \left\{ \frac{\mu_e \tau_e E}{D} \left[1 - \exp\left(\frac{x_i - D}{\mu_e \tau_e E} \right) \right] + \frac{\mu_h \tau_h E}{D} \left[1 - \exp\left(\frac{-x_i}{\mu_h \tau_h E} \right) \right] \right\},$$ (3)

and for a planar detector, the electric field intensity will be $E = V/D$, where V is the applied voltage.

In practice, when using the Hecht equation to determine the $\mu\tau$-product, charge carriers must be generated close to either surface of the detector, for example, using a laser pulse, α particles, or a low energy X-ray source. This will result in the signal mainly depending on one carrier, and the one-charge Hecht equation can be used (excluding the part of the equation describing the other carrier). The CCE of the detector is then determined, given by

$$\text{CCE} = \frac{Q}{Q_0} = \frac{Q}{eN_0},$$ (4)

where Q_0 is the induced charge if the initial number of generated electron–hole pairs are all collected. A set of measurements are then taken operating the detector at different bias voltages, thus varying the electric field strength inside the detector. The variation in the induced charge Q (measured by photopeak amplitudes) is registered for each bias setting. Next, the $\mu\tau$-product is determined by plotting CCE as a function of applied bias and then fitting the Hecht equation using a curve fitting procedure. Various methods exist for determining Q_0, for example, by taking a measurement maximum bias, thus minimizing trapping much as possible, or for another method, data originating very close to the collecting surface is used, thus minimizing amount of trapping.

Determining $\mu\tau$-products using the Hecht equation is one of the most conventional methods used, has been used by many research groups and crystal manufactures, and is often used for comparative testing of CZT materials. Commonly the $\mu\tau$-products are determined as a mean value for the entire detector bulk [9]. Examples have been seen for mapping the $\mu\tau$-product spatially, for example, [19] mapped $\mu_e\tau_e$ of CZT detectors using a microbeam by determining CCE distribution spatially and at different bias voltages. Thus, for CCE measurement of the detector, $\mu_e\tau_e$ was extracted by fitting the Hecht equation using a Levenberg–Marqardt algorithm.

The precision of the $\mu\tau$-estimation using the Hecht equation is affected by different factors. First of all, a curve fitting procedure is necessary to determine $\mu\tau$. Second, the method tends to underestimate $\mu\tau$ due to ballistic deficit and surface trapping [3, 12]. Furthermore, the Hecht equation assumes the charge trapping to be uniform within the detector material, which does not take into account the possibility of non-uniformity of the detector material, unless specific methods are used to spatially map the product, as done by Lohstroh et al. [19]. The Hecht equation assumes a constant electric field inside the detector, which thereby also assumes the drift time to be linear function of distance. At low biases, this is often not fulfilled, resulting in measurements being taken at higher bias. This often also results in underestimated values of $\mu\tau$, especially for high $\mu\tau$-product materials [2].

2.2   Direct Measurement Method

He et al. [12] suggested two new approaches to simply calculating the $\mu_e\tau_e$-product for electrons using single polarity charge sensing depth sensing CZT detector [13, 14, 20]. The two direct methods presented for determining the $\mu_e\tau_e$-product are, firstly, measuring photopeak amplitudes versus the drift length and secondly measuring photopeak amplitudes versus the mean drift length. The two methods first described by He et al. [12] will be summarized here. Both methods require the detector to be single polarity charge sensing.

The first method assumes the fractional electron loss *(dN/N)* per unit path length to be constant and requires the depth of interaction to be determined. The exponential relationship describing the decay of electrons with a given drift distance y_d is given by

$$N = N_0 \exp\left(\frac{-y_d}{\mu_e\tau_e E}\right), \tag{5}$$

where N is the number of electrons collected after a drift distance y_d and N_0 is the original number of electron–hole pairs generated. Two measurements can then be extracted, one for events close to the cathode surface and one for events close to the anode surface. Ignoring the electron loss for gamma-rays interacting close to the anode surface, the difference in photopeak amplitudes (assuming the amplitudes are proportional to the number of collected electrons) between the two measurements spectra will show the electron lost due to drifting from the cathode side of the detector to the anode side. Thus, letting N_0 be the photopeak amplitude of events originating close to the anode surface and N be the photopeak amplitude from events originating near the cathode surface, then the drift distance y_d will be equal to the detector thickness D, we rewrite Eq. (5) such that the $\mu_e \tau_e$-product is given by

$$\mu_e \tau_e = -\frac{D}{\ln(N/N_0)E}. \tag{6}$$

This method allows for the usage of a high-energy X-ray source, since the depth sensing technique takes care of extracting the photon interaction positions and distinguishing events close to the anode from the ones close to the cathode. The issue of this method is that for many electron-only CZT detectors the electrode configuration is such that the region close to the anode is nonlinear. This can introduce a systematic error when determining the photopeak amplitude of events originating close to the anode and thus introducing an error assuming the photopeak amplitude of events in this region being a measure of N_0. So, even though this technique overcomes some of the known issues of determining $\mu\tau$ using the Hecht equation, the systematic error of assuming N_0 incorrectly can introduce a wrong determination of $\mu_e \tau_e$

He et al. [12] introduced another technique, reducing systematic errors when estimating $\mu_e \tau_e$. This method does not require the determination of N_0, contrary to the previous method. In this technique, two photopeak amplitudes (N_1 and N_2) are measured considering events close to the cathode surface at two bias voltages V_1 and V_2 between cathode and anode. Consequently, this also means measurements at two different electric field strengths (E_1 and E_2) and thereby also two different measured photopeak amplitudes (N_1 and N_2) rewriting Eq. (5),

$$N_1 = N_0 \exp\left(-\frac{y_d}{\mu_e \tau_e E_1}\right)$$

$$N_2 = N_0 \exp\left(-\frac{y_d}{\mu_e \tau_e E_2}\right). \tag{7}$$

Solving allows us to become independent of the initial number of electron–hole pairs, N_0, such that

$$\mu_e \tau_e = \frac{y_d}{\ln(N_1/N_2)}\left(\frac{1}{E_2} - \frac{1}{E_1}\right). \tag{8}$$

At lower bias voltage, the electric field strength will be weaker and thereby the mean free path of the electrons shorter. Therefore, the photopeak amplitude will be lower for data acquired at lower bias voltage, when considering events with the same drift length. He et al. [12] showed that their estimates of $\mu_e \tau_e$ using these two methods yielded larger values compared to what was measured using the conventional Hecht equation method. The techniques can also be applied for determining $\mu_h \tau_h$-product of the holes but are limited by the single polarity sensing requirement. Furthermore, since the depth of interaction is known, energy spectra for events originating near cathode surface can instead be evaluated at various depths throughout the detector material.

The method requires two bias measurements, but it can be expanded to a set of bias measurements. Thus, by fixing the highest bias measurement V_2 as reference measurement, extracting N_2, several measurements can be used to determine different values of N_i, at different bias settings V_i. A set of points can be calculated using Eq. (9), plotting $\ln(N_i/N_2)$ as a function of $(1/E_2 - 1/E_i)$ and fitting a linear regression model with the slope, where i indicates the index number of the bias measurement

$$\ln\left(\frac{N_i}{N_2}\right) = \frac{y_d}{\mu_e \tau_e}\left(\frac{1}{E_2} - \frac{1}{E_i}\right). \tag{9}$$

The slope will then be the relation

$$\alpha = \frac{y_d}{\mu_e \tau_e}. \tag{10}$$

By knowing the electron drift length, $\mu_e \tau_e$ can be estimated from a set of measurements at different bias [21].

With 3D position sensitivity of the detectors, it is possible to map the variation of material characteristics throughout a detector volume, which was done for pixelated 3D CZT detectors by He et al. [15], estimating the mean $\mu_e \tau_e$-product underneath each anode pixel. It has been implemented, calculating a 3D map distribution of $\mu_e \tau_e$ [21], which will be presented in a later section of this chapter.

2.3 Drift Time Method

Bolotnikov et al. [2] introduced the drift time method for determining the electron lifetime τ_e alone, a method that is independent of the electric field distribution in the crystal and eliminates ballistic deficit effects. The method is based on measuring as long as possible drift times for electrons using thick detectors together with low biases. This method is especially suited for crystals where the electron charge loss across the detector is small. In this method, events originating close to the cathode were considered.

We consider the equation describing electron decay time, this time in relation of electron drift time t_d

$$N = N_0 \exp\left(-\frac{t_d}{\tau_e}\right). \tag{11}$$

Bolotnikov et al. [2] evaluates N at lowest possible bias and N_0 at highest possible bias. Using this and rewriting Eq. (11), the electron lifetime can then be calculated by

$$\tau_e = \frac{t_d}{\ln\left(\dfrac{N_0}{N}\right)}. \tag{12}$$

Bolotnikov et al. [2] found that this method estimated $\mu_e \tau_e$-values one order of magnitude larger than when determined using the Hecht equation.

This method can similar to the previously presented method be extended to include more than two bias measurements and thus becoming independent of the N_0 measurement [21]. This makes use of Eq. (11) resulting in the expressions, where they are measured at two different bias voltages, thus resulting in two different drift times t_{d1} and t_{d2}

$$N_1 = N_0 \exp\left(-\frac{t_{d1}}{\tau}\right)$$
$$N_2 = N_0 \exp\left(-\frac{t_{d2}}{\tau}\right). \tag{13}$$

Solving these two equations allows us to become independent N_0, such that

$$\ln\left(\frac{N_1}{N_2}\right) = \frac{1}{\tau}(t_{d2} - t_{d1}), \tag{14}$$

which requires only considering data taken at two bias measurements. This method can similar to $\mu\tau$ be expanded such that the highest bias measurement is used as reference measurement, extracting N_2. Thus, several measurements can be used to determine different values of N_1, at different bias settings, yielding N_i for i measurements. A set of points can be calculated using Eq. (14), plotting $\ln(N_i/N_2)$ as a function of $(t_{d2} - t_{di})$ and fitting a linear regression model with the slope

$$\alpha = \frac{1}{\tau}. \tag{15}$$

2.4 Time of Flight

A general transient current or charge technique used to determine μ of the semiconductor detector bulk is the time of flight (TOF) method [27, 28]. In this method, the current or voltage pulse shapes are analyzed to obtain information on the drift time of the charge carriers [1]. For this method, it is important that the bias is increased such that the drift time t_d is much larger than the charge carrier lifetime τ [10].

Different approaches have been taken when using this method. The detector electrode surface can be irradiated by a shallow penetrating radiation such as alpha particles [9, 29, 35] or a pulsed laser light [7, 10, 31, 34]. If the detector is depth-sensitive, a high-energy X-ray source can be used and events at a given distance to the electrode surface extracted for the analysis.

Often a mean value for the entire detector bulk is determined. Spatial uniformity of the material parameter can also be investigated with different approaches. Sellin et al. [29] used ion beam induced charge (IBIC) to map the mobility spatially. Position information of events will allow the same mapping.

In the TOF method, the charge carriers will drift toward collection under the influence of an applied bias, V, generating an electric field, E, inside the semiconductor detector material. The induced charge generated by the drifting charge carrier is sampled, and the drift time (pulse rise time) can from this be extracted. The speed of the carrier is given by

$$v = \frac{y_d}{t_d}, \tag{16}$$

where y_d is the drift distance [27]. For events originating close to the non-collecting electrode surface, this will be equal to the thickness of the detector, D. Using this, the mobility can then be calculated by

$$\mu = \frac{v}{E} = \frac{y_d}{E t_d}. \tag{17}$$

3   3D Mapping of Mobility and Lifetime in CZT

In the previous sections, some different approaches used for determining the $\mu\tau$-product of CZT material were summarized. In the following sections, we will present the 3D mapping of μ_e and τ_e utilizing a CZT drift strip detector. These results were initially presented in [21]. The detector used for the analysis is the 3D CZT drift strip detector developed at DTU Space [18, 23, 24]. The detector is single polarity sensing due to the specific design of the electrode configuration and biasing. The design is specifically intended to overcome the ineffective charge collection in CZT due to the poor hole mobility. The collecting anodes of the detector are screened from the hole movement, resulting in the detector being an electron-only device, which as an example also was a requirement for the $\mu_e \tau_e$ extraction methods proposed by He et al. [12]. Describing the variations in μ_e and τ_e inside the bulk instead of a mean value will provide an understanding of given detector material properties and make us able to get a more coherent understanding on the signal formation in the detector. This will not only describe the non-uniformity of the detector material but can also be used as look-up tables for trapping, which will greatly improve the energy resolution. Furthermore, it can be implemented in a detector model to better predict electron movement characteristics in the detector and by that better predict pulse shape formation [21].

In the following sections, a short introduction to the specific detector will be given, followed by a summary of the experimental setup used for the specific $\mu_e \tau_e$ extraction methods used. Lastly, we will summarize the data analysis and $\mu_e \tau_e$ calculations performed resulting in a 3D spatial map of the material properties for the electrons, illustrating a 3D implementation of some common $\mu_e \tau_e$ extraction methods.

3.1 Detector Operation

A short description of the detector used for this specific $\mu_e \tau_e$-extraction study will be given here, and further information on the 3D CZT drift strip detector can be found in the following works [5, 18, 21–24]. In Fig. 1a, a schematic of the 3D CZT drift strip detector is shown, and Fig. 1b shows an image from the anode side of the detector. The detector crystal is of size 2 cm × 2 cm × 0.5 cm. On one side of the detector, 12 anodes (red) are deposited, and separating these are bias drift strip sections (blue). Each drift strip section consists of three drift electrodes, except for the two outer sections which consist of two electrodes. On the other side of the detector crystal, 10 cathode strips are deposited perpendicular to the anodes and drift strips [18]. For standard operation for the given prototype used, the anodes are grounded and the cathodes are biased at −350 V. The drift strips are biased at a drift voltage of −120 V, a voltage divider results in the central drift strip electrodes being held at a bias of 2/3 of the drift voltage, and the outer adjacent drift strips are held at 1/3 of the drift voltage. With the applied bias, the electric field lines are concentrated toward the anodes for electron collection. Figure 2 illustrates the detector from the XY-plane. The electric field lines are concentrated toward the collecting anodes due to the bias configuration of the detector. Furthermore, 12 drift sections exist, one for each anode. A generated electron charge cloud from a photon interaction occurring in a given drift section will be collected at the anode in the center of the section.

For readout, bi-polar charge-sensitive pre-amplifiers are used. The output is sampled by high-speed digitizers, with a sampling rate of 4 nanoseconds. All anodes and cathodes are read out in separate channels, and the drift strips are sectioned into four readout sections. The final readout of the 3D CZT drift strip detector therefore consists of 26 pulse shapes. An example of such 26 pulse shapes, resulting

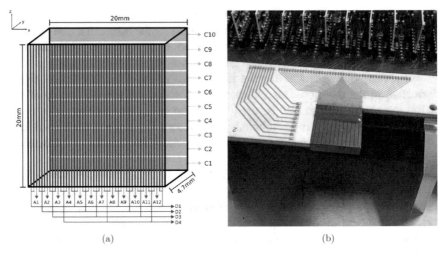

(a)   (b)

Fig. 1 Schematic and image of the 3D CZT drift strip detector. (**a**) Schematic. (**b**) Real detector

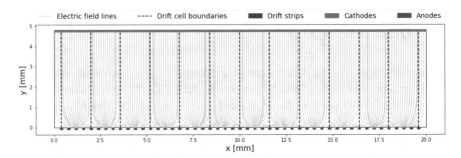

Fig. 2 Illustration of the electric field line distribution in the 3D CZT drift strip detector calculated by COMSOL at standard operating bias configuration

from a photon interaction with the detector material, can be seen in Fig. 3. These pulse shapes are a unique fingerprint holding information on what went on in the detector. Analyzing the shapes of these transient charge signals can provide information on the photon interaction position inside the detector in 3D. It can provide information on the energy deposited by the photon, the drift time of the electrons, and furthermore also tell what kind of event took place (photoelectric absorption, Compton scattering, or pair production). The specific algorithms used to extract these information from the transient charge signals can be found in [5, 21, 22]. The pulse shapes are very distinctive, due to the geometry of the electrodes, and thereby the resulting weighting potentials [11, 25, 30]. Figure 4 illustrates weighting field distribution in the detector for a subsection of the detector simulated using COMSOL. The left most figure shows the weighting potential for the anode, the center figure the weighting potential for a cathode, and lastly to the right, the weighting potential for the drift strip section to the right of the anode. The

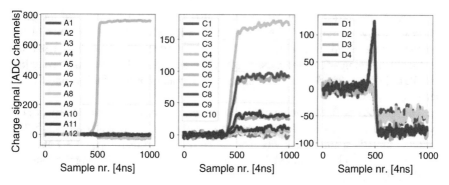

Fig. 3 Pulse shapes of a given event generated in the 3D CZT drift strip detector, for anodes (left), cathodes (center), and drift strips (right)

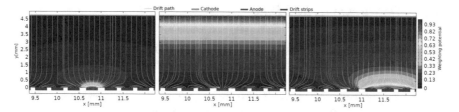

Fig. 4 COMSOL illustration of weighting potential distribution in the 3D CZT drift strip detector, for an anode (left), cathode (center), and drift strip section (right)

overall pulse shape appearance depends on the charge cloud movement through the weighting potentials. For example, when inspecting the anode signal in Fig. 3(left), we see that the anode signal for anode number 7 experiences a sudden steep rise, while the neighboring anodes do not sense the charge movement, due to the compact weighting potential around the anode electrode, as seen in Fig. 4(left). At collection, the anode signal flattens, since no charge movement is inducing signal. The cathode signal in Fig. 3(center) shows several of the cathodes sensing the charge movement, from the moment the charges starts moving increasing linearly, due to the distribution of the weighting potential for the cathodes, seen in Fig. 4(center). At collection, the cathode signal flattens but still has a small increasing slope. This is due to the holes still moving inducing some signal. The drift strip signal seen in Fig. 3(right) has a distinctive shape increasing followed by a sudden decrease. This can be described by inspecting the weighting potential distribution for the drift strip section, Fig. 4(right). The specific pulse shapes see the charge cloud moving toward collection, thereby increasing the drift strip signal. Just before collection, the charge cloud moves away from the drift strip section toward the collecting anode, thereby moving out of the weighting potential related to the given drift strip section, resulting in a sudden drop in the pulse shape. Predicting these pulse shapes can, as mentioned earlier, provide information on what went on in the detector, but the weighting potential is not the only parameter affecting the pulse shape formation.

Material properties of the detector will also affect the appearance of the pulse shapes. Firstly, charge drift time is strongly affected by the $\mu\tau$-product of the material, since the speed at which the charge moves affects the pulse shape rise time. Secondly, trapping affects the pulse heights. For an almost electron-only device like the 3D CZT drift strip detector, especially $\mu_e\tau_e$ has an impact on the generated pulse shapes.

The 3D CZT drift strip detector has previously demonstrated both excellent 3D - position (<0.5 mm at 661.6 keV) and energy resolution ($\sim 1\%$ at 661.6 keV) [5, 22], and it is this 3D spatial sensitivity that can be utilized to map $\mu_e\tau_e$ in 3D and thus investigate the uniformity of the detector crystal itself. In the following sections, the spatial extraction of $\mu_e\tau_e$ will be described, initially presented in [21].

3.2 Measurements and Data

Interpreting the distinct features of the 26 pulse shape signals for each photon inter-action in the 3D CZT drift strip detector, information can be extracted and utilized to determine the $\mu_e\tau_e$-product. Equations (9) and (14) were used to determine $\mu_e\tau_e$ and τ_e, respectively. When estimating these parameters, measurements must be taken varying the bias between anode and cathode, and the detector must return deposited energy, electron drift time, and electron drift distance. For the spatial mapping of the material properties, the detector should also return 3D interaction position information for each event.

For the experimental setup, a ^{137}Cs source was used fully illuminating the XZ-plane of the detector. In this way, photon interactions occur in the entire detector material. Twenty hour measurements were taken at each bias setting, varying the cathode bias from -150 to -400 V with a voltage increase between each step of 50 V, resulting in six bias measurements. The drift bias was set to a third of the cathode bias. Each measurement for a given bias results in a data set, for which photoelectric absorption events are extracted (single hit events in the photopeak). Pulse shape analysis is then performed to extract pulse heights, 3D position information, and drift times. The electron drift distance can then be calculated knowing the position information together with the electron drift time [21]. The detector is divided into voxels, and data is assigned to a voxel, given the 3D position information of the event. The $\mu_e\tau_e$ analysis is then performed for each voxel, using the data assigned.

For this specific analysis, a subset of the detector volume was analyzed, but the entire detector volume can be included. The subset region chosen is highlighted in the 2-D position histogram of a data set displayed in Fig. 5. The subset region for which the 3D maps of electron mobility and lifetime are extracted consists of anodes 2–11. Furthermore, the y-position of the region is restricted to be 2.0 mm $\leq y \leq$ 4.0 mm, and lastly the z-position region is restricted to be 13.0 mm $\leq z \leq 15.0$ mm. This limitation is done due to some of the artifacts which has also been earlier encountered. First of all, we see no events for $z < 2$ mm and $z > 18$ mm.

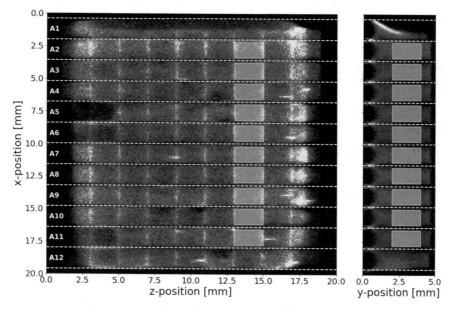

Fig. 5 2D histograms displaying the calculated 3D positions for the measurement taken with a cathode bias of $-350\,$V and a drift bias of $-120\,$V. White dashed lines indicate drift cell boundaries. Anode numbers are defined within each drift cell. The white highlighted areas indicate the chosen data subset of the detector volume. © 2021 IEEE. Reprinted, with permission, from [21]

This is a severe edge effect following the current implementation of z-position determination [5, 21, 22]. Algorithm adjustments to overcome this issue are in the pipeline. Furthermore, we have some dead-zone areas where we have an attenuation on photon interactions or no signal at all (around anodes 5, 6, 11, and 12 (A12) for $z < 5.0\,$mm). This has also earlier been encountered [22] and is an artifact for this specific detector prototype. For the histogram displaying positions of the XY-plane, we also see artifacts earlier encountered [5, 22], where the edge anodes show clear edge effects together with the current x-position algorithm not working for events where $y < 1\,$mm. Along the x-direction, the region is further restricted, since only data 0.2 mm from the drift cell boundary is considered, since shared events between anodes are not included in the analysis. The white highlighted areas in Fig. 5 give a quick indication of the data subset used for this analysis [21].

The 3D map itself is a voxel division of this specific subset region. The 3D map voxel division consists of voxels of size $(dx, dy, dz) = (0.4\,$mm, $0.2\,$mm, $1.0\,$mm$)$. The size is mainly limited by the statistics, since enough events within each virtual volume are needed for the given exposure time of the measurements. dy was chosen to be minimized as much as possible, since the drift time is greatly affected by the interaction depth, and thus the larger a range of y-values on each virtual volume, the greater spread in the drift time values for the given volume. Table 1 gives an overview of the data division used for generating the 3D maps.

Table 1 Overview of the 3D data division boundaries. © 2021 IEEE. Reprinted, with permission, from [21]

Direction	Number of slices	Slice thickness [mm]	Minimum value [mm]	Maximum value [mm]
x (anode 2)			2.2	3.4
x (anode 3)			3.8	5.0
x (anode 4)			5.4	6.6
x (anode 5)			7.0	8.2
x (anode 6)	3	0.4	8.6	9.8
x (anode 7)			10.2	11.4
x (anode 8)			11.8	13.0
x (anode 9)			13.4	14.6
x (anode 10)			15.0	16.2
x (anode 11)			16.6	17.8
y	10	0.2	2.0	4.0
z	2	1.0	13.0	15.0

The resulting 3D maps are presented in the following sections. All error estimation is calculated through error propagation, assuming the electric field strength is known. For all bias settings, the electric field strength was extracted using COMSOL [21].

3.3 $\mu_e\tau_e$-Product

To map the $\mu_e\tau_e$-product in the 3D CZT drift strip detector measurements were taken at six different bias settings, and Eq. (9) is applied. The photopeak amplitude N_2 is described by the measurement taken at the maximum bias voltage of -400 V for the cathodes. N_i is described by the photopeak amplitudes for the remaining five bias measurements. For each voxel in the subsection of the detector volume, photopeak amplitude and mean drift time must be extracted, for each single bias measurement. Thus, for each bias measurement, 600 photopeak amplitudes and drift times are extracted. Figure 6 illustrates the photopeak fitting procedure for three voxels in the detector. The photopeak of data inside each voxel is fitted with a Gaussian and the peak position extracted. The same procedure is done for electron drift times, resulting in each voxel (for each bias measurement,) having assigned a photopeak amplitude and an electron drift time. The electron drift distance for the given voxel is then extracted, knowing both y-position and electron drift time [21]. For a planar detector, the drift distance is equal to the interaction depth (y-position), since the electric field lines, and thus the drift paths are linear and perpendicular to the anode. For the 3D CZT drift strip detector, this is not the case, which was also demonstrated in Fig. 2. Actually, the y-position in the 3D CZT drift strip detector is only equal to the electron drift length, in the case where the absorption occurred

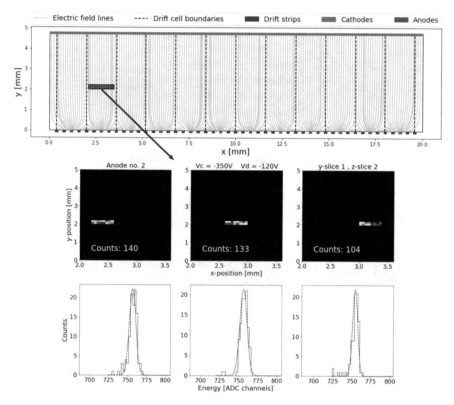

Fig. 6 Illustration of photopeak fitting for three voxels in the 3D CZT drift strip detector

directly above the anode in the center of a drift cell. For each voxel in the 3D sectioning of the detector, the drift distance must be estimated, since events closer to drift cell boundaries (above a drift section) will experience a longer drift path and therefore also more trapping. The drift distance was determined by calculating the electron velocity for a data voxel right above the anode where the drift distance is known and then dividing drift distance with drift time. Thus, by knowing the drift velocity, the drift distance was calculated for the neighboring voxels to the central voxel for each anode section; for further information, see [21].

For each voxel, $\ln(N_i/N_2)$ as a function of $(1/E_2 - 1/E_i)$ was plotted and fitted with a linear regression; see an example fit of three voxels in Fig. 7. N_i and E_i represent values from the first five measurements and N_2 and E_2 from the highest bias measurements. Electric field values were extracted from COMSOL models of the detector. Each fit contains five points, for each of the 600 voxels; thus 600 linear regressions were fitted, and the slope extracted and used to estimate the $\mu_e \tau_e$-product of each voxel. Figure 8 displays the extracted 3D map of the $\mu_e \tau_e$-product. A clear inhomogeneity is visibly with values increasing along the x-direction of the detector. Actually, the $\mu_e \tau_e$-product values are varying throughout the material with

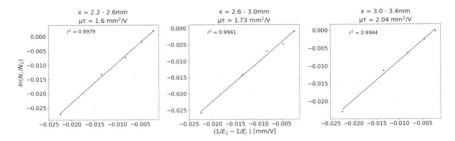

Fig. 7 Example of the $\mu_e\tau_e$-product extraction using a linear fit for three voxels

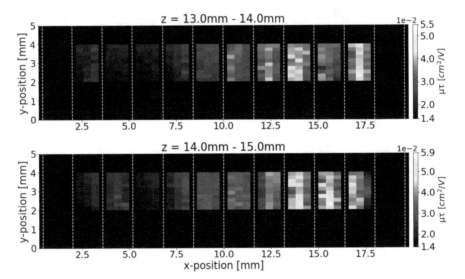

Fig. 8 $\mu_e\tau_e$-map of the chosen detector sub-volume. The mean estimated error of the calculated $\mu_e\tau_e$-values is ~10%. White dashed lines: drift cell boundaries. © 2021 IEEE. Reprinted, with permission, from [21]

a factor of almost three. The estimated error determined through error propagation showed an uncertainty of ~10%, and the variation in $\mu_e\tau_e$-product is greater than this error [21].

3.4 τ_e

Similar to the $\mu_e\tau_e$-product, the electron drift time and photopeak positions were extracted from all 600 voxels for the six data set taken at different bias. In the case of extracting τ_e, no electric field or drift distance was needed, and the linear fit, $\ln(N_i/N_2)$ as a function of $(t_{d2} - t_{di})$, was fitted extracting the slope to determine a value of τ_e, one for each voxel. A similar approach was taken, fixing N_2 and t_2 to be

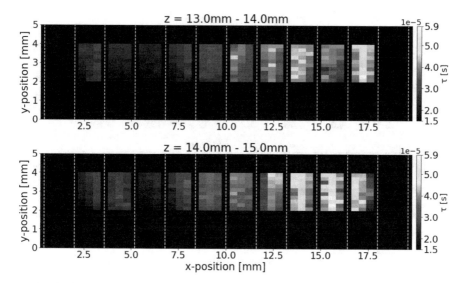

Fig. 9 τ_e-map of the chosen detector sub-volume. The mean estimated error of the calculated τ_e-values is $\sim 5\%$. White dashed lines: drift cell boundaries. © 2021 IEEE. Reprinted, with permission, from [21]

the measurement taken at highest bias and N_i and t_{di} being the measurements taken at the remaining bias settings. This results in five points plotted and fitted with a linear regression for each voxel, using the slope to extract τ_e. The resulting 3D map is seen in Fig. 9, where similar tendencies as for $\mu_e\tau_e$-map are seen. The error was again determined by error propagation to be $\sim5\%$. Since the variation in $\mu_e\tau_e$ and τ_e is so similar, it seems electron lifetime is the main contributor to the inhomogeneity.

3.5 μ_e

μ_e was extracted using the values for $\mu_e\tau_e$ and τ_e. The resulting 3D map is seen in Fig. 10. Error propagation was determined to $\sim10\%$. Variances in the map are smaller compared to the previous maps. It would be possible to implement the TOF method for each voxel, thereby calculate μ_e independently on the two previous methods, and compare to this result.

3.6 Modeling Detector Response

Finally, evaluating the extracted $\mu_e\tau_e$ 3D maps, they can be implemented in a model for the detector, to evaluate the detector performance on predicting detector pulse

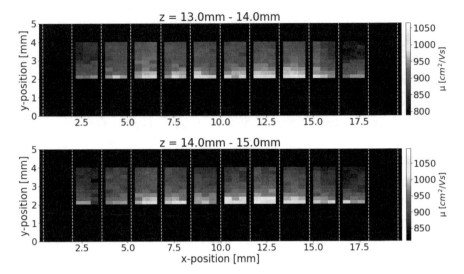

Fig. 10 μ_e-map of the chosen detector sub-volume. The mean estimated error of the calculated μ_e-values is $\sim$ 10%. White dashed lines: drift cell boundaries. © 2021 IEEE. Reprinted, with permission, from [21]

shapes. In the final two sections of this chapter, we will introduce a model of the 3D CZT drift strip detector and see how it predicts the pulse shapes using the newly calculated 3D maps [21] (Figs. 8, 9, and 10).

3.6.1 COMSOL Model

The pulse shapes generated by charge collection in the detector contain a lot of information on how the photon interacted in the detector and how the generated charge carriers drifted toward collection at the anode. The pulse shape formation is also used to describe the 3D position of the photon interaction, the deposited energy, and charge trapping over the detector volume. If a trustworthy model of the 3D CZT drift strip detector can be developed, pulse shape formation of any event can be predicted. A model that reliably predicts pulse shape formation in the detector can be used to generate training data for artificial neural networks.

With a combined use of COMSOL Multiphysics® and Python, a model of the 3D CZT drift strip detector has been developed [21]. The 3D CZT drift strip detector geometry was implemented in COMSOL and electrostatic conditions of the detector calculated, such that weighting potential and electric field strengths can be extracted. The Python model takes this information as input together with sample time and length, start time, and initial 3D position of the photon interaction. Python will then in an iterative calculation calculate induced signal, charge carrier movement, and charge trapping with a sample rate of 4 nanoseconds similar to the real detector setup, until the sample time is reached. For each iteration, the induced signal is calculated utilizing the weighting potential information extracted by COMSOL. Next, the charge positions are updated for both charge carriers, by calculating the change in position dL for a sample time of t_s, by

$$dL = \mu E t_s \qquad (18)$$

utilizing the electric field information E supplied by COMSOL and the mobility of the charge carrier μ in the CZT material provided by a 3D map. Next, the electron trapping information is updated first by calculating the free drift length of the charge carrier by

$$\lambda = \mu \tau E \qquad (19)$$

and then calculating trapping by the exponential relationship

$$N_{dL} = N \exp\left(-\frac{dL}{\mu \tau E}\right), \qquad (20)$$

where N_{dL} is the number of charge carriers after the iterative step and N is the number of charge carriers before the movement of dL. After the sample time is reached and thereby all iterations finished, the model outputs a similar format as for the true detector system, with 26 pulse shapes generated for each interaction. These calculations performed by the detector all depend on the charge carrier transport properties μ and τ. Therefore, for the model to be trustworthy, these properties of the detector material must be well understood. The induced signals are especially affected by the $\mu\tau$-values of the electrons, since the detector is close to an electron-only device. Furthermore, the detector material is not necessarily homogeneous, and therefore, the variation in $\mu\tau$ over the detector material is important to quantify, for the model to create truly reliable predictions of the generated pulse shapes. Therefore, the model is expanded to take the mobility lifetime of the electrons as an input, and the extracted 3D map used as look-up table by the model [21].

3.6.2   Implementation of $\mu_e \tau_e$-map

The 3D $\mu_e \tau_e$-map is implemented in the 3D CZT drift strip detector model such that the model looks up the μ_e and τ_e values corresponding to the voxel within the event took place. Figure 11a shows the pulse shapes generated by a single event interaction in the detector. From the detector data, position and start time were calculated and used as input for the model. The model then simulates the event starting at the given start time and location. Figure 11b shows the resulting electron–hole trace of the interaction during the sampling time. Here, we see the electrons drift toward collecting at the anode in the second drift section and the holes drift toward collection at the cathode. However, they are not collected during the sampling time due to the poor mobility. For this given event, the μ_e and τ_e values

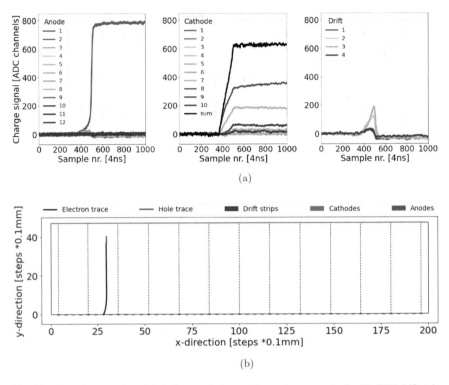

Fig. 11 Pulse shapes and modeled electron–hole trace for a given event in the 3D CZT drift strip detector. (**a**) Pulse shape response of a photon interaction in the detector. Calculated position: $(x, y, z) = (2.97\text{mm}, 3.70\text{mm}, 14.50\text{mm})$ and $t_0 = 364.27\text{s}$. (**b**) Model response of electron–hole trace

were extracted from the 3D maps, returning the values $\mu_e = 858.93\,\text{cm}^2/\text{Vs}$ and $\tau_e = 2.5 \cdot 10^{-5}\text{s}$. Figure 12 shows a comparison of the generated pulse shapes from the real event, together with the simulated pulse shapes returned by the model, using the generated 3D maps. In this figure, the main affected electrodes are shown, but the model does return all 26 pulse shapes. Inspecting the figure we see for the three anode signals (Fig. 12a–c), we see a clear compliance between model and data. The rise time of the triggered anode is nicely in agreement with the modeled data, and the two neighboring anodes show a small increase in signal followed by a decrease around collection, also verified by the model. We do see for the collecting anode (Fig. 12b) that the model has a very clear collection where the signal flattens, where for the real data, this area is more rounded. This is due to the model simulating a single point charge, but the real pulse shape is generated by a charge cloud. For the cathode signals (Fig. 12d–f), we again see compliance between model and data for the pulse rise time for the central cathode and for the neighboring. However, we do see discrepancy after the charge cloud has been collected, where for the model pulse shapes, the signal flattens fully. However, for the cathode signal of the real

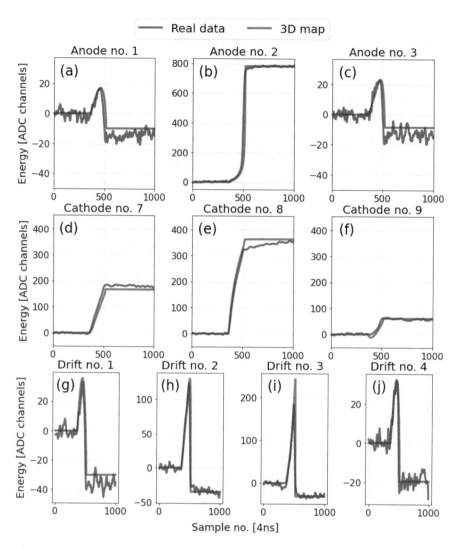

Fig. 12 Comparison between pulse shapes of a real event and pulse shapes simulated by the model using the μ_e and τ_e values extracted from the calculated 3D maps

pulse shapes, after collection, we see a small increasing slope in signal. This is due to hole movement in the detector, indicating better hole mobility and lifetime than initially expected. Current fixed values for hole mobility and lifetime in the detector model are set to $\mu_h = 20\,\text{cm}^2/\text{Vs}$ and $\tau_h = 3 \cdot 10^{-7}\text{s}$. This invites for an investigation of hole mobility and lifetime, for more correct prediction of pulse shapes by the model. For the drift strip signals (Fig. 12g–j), once again we see a good compliance between model and data using 3D maps of material characteristics. Currently, the model uses discrete values for each given voxel in the 3D map, and however, in reality $\mu_e \tau_e$ are not discrete but vary gradually throughout the material. Therefore, interpolation of the 3D maps should be implemented [21].

4   Final Remarks

In this chapter, we have introduced several methods which can be used for determining μ and τ for CZT detectors. The method applied and the source used depend on the detector and its capabilities. We showed that for a 3D position-sensitive CZT detector, some of the methods could be utilized to map the 3D distributions of the material properties in the detector bulk. Thus, 3D maps of electron lifetime and mobility can be extracted and implemented in a detector model, for more precise pulse shape prediction. Modeled pulse shapes compared to the pulse shapes of the real event demonstrated compliance between predicted and true pulse shapes. Furthermore, discrepancy between model and real data was shown in the hole properties for the material being under-estimated, inviting for further investigation of these values. Understanding the material properties and their inhomogeneity in 3D is important not only for providing better detector model predictions but can also be used for data correction. A model correctly simulating the pulse shapes will also be an excellent candidate to generate training data for artificial neural networks [21].

References

1. Bell, R., Wald, F., Canali, C., Nava, F., & Ottaviani, G. (1974). Characterization of the transport properties of halogen-doped CdTe used for gamma-ray detectors. *IEEE Transactions on Nuclear Science, 21*(1), 331–341.
2. Bolotnikov, A. E., Camarda, G. S., Chen, E., Gul, R., Dedic, V., Geronimo, G. D., Fried, J., Hossain, A., MacKenzie, J. M., Ocampo, L., Sellin, P., Taherion, S., Vernon, E., Yang, G., El-Hanany, U., & James, R. B. (2016). Use of the drift-time method to measure the electron lifetime in long-drift-length CdZnTe detectors. *Journal of Applied Physics, 120*(10), 104507.
3. Boucher, Y. A., Zhang, F., Kaye, W., & He, Z. (2012). New measurement technique for the product of the electron mobility and mean free drift time for pixelated semiconductor detectors. *Nuclear Instruments and Methods in Physics Research Section A: Accelerators, Spectrometers, Detectors and Associated Equipment, 671*, 1–5.
4. Budtz-Jorgensen, C., Kuvvetli, I., Skogseide, Y., Ullaland, K., & Ostgaard, N. (2009). Characterization of CZT detectors for the ASIM mission. *IEEE Transactions on Nuclear Science, 56*(4), 1842–1847.
5. Budtz-Jørgensen, C., & Kuvvetli, I. (2017). New position algorithms for the 3D CZT drift detector. *IEEE Transactions on Nuclear Science, 64*(6), 1611–1618.
6. Burrows, D. N., Hill, J. E., Nousek, J., Kennea, J. A., Wells, A., Osborne, J. P., & Hartner, G. D. (2005). The swift x-ray telescope. *Space Science Reviews, 120*(3), 165–195.
7. Burshtein, Z., Jayatirtha, H., Burger, A., Butler, J., Apotovsky, B., & Doty, F. (1993). Charge-carrier mobilities in Cd0. 8Zn0. 2Te single crystals used as nuclear radiation detectors. *Applied Physics Letters, 63*(1), 102–104.

8. Eisen, Y. (1996). Current state-of-the-art industrial and research applications using room-temperature CdTe and CdZnTe solid state detectors. *Nuclear Instruments and Methods in Physics Research Section A: Accelerators, Spectrometers, Detectors and Associated Equipment, 380*(1–2), 431–439.
9. Eisen, Y., Shor, A., & Mardor, I. (1999). CdTe and CdZnTe gamma ray detectors for medical and industrial imaging systems. *Nuclear Instruments and Methods in Physics Research Section A: Accelerators, Spectrometers, Detectors and Associated Equipment, 428*(1), 158–170.
10. Erickson, J., Yao, H., James, R., Hermon, H., & Greaves, M. (2000). Time of flight experimental studies of CdZnTe radiation detectors. *Journal of Electronic Materials, 29*(6), 699–703.
11. He, Z. (2001). Review of the Shockley–Ramo theorem and its application in semiconductor gamma-ray detectors. *Nuclear Instruments and Methods in Physics Research Section A: Accelerators, Spectrometers, Detectors and Associated Equipment, 463*(1–2), 250–267.
12. He, Z., Knoll, F., & Wehe, D. K. (1998). Direct measurement of product of the electron mobility and mean free drift time of CdZnTe semiconductors using position sensitive single polarity charge sensing detectors. *Journal of Applied Physics, 94*(10), 5566–5569.
13. He, Z., Knoll, G., Wehe, D., & Miyamoto, J. (1997). Position-sensitive single carrier CdZnTe detectors. *Nuclear Instruments and Methods in Physics Research Section A: Accelerators, Spectrometers, Detectors and Associated Equipment, 388*(1–2), 180–185.
14. He, Z., Knoll, G. F., Wehe, D. K., Rojeski, R., Mastrangelo, C. H., Hammig, M., Barrett, C., & Uritani, A. (1996). 1-D position sensitive single carrier semiconductor detectors. *Nuclear Instruments and Methods in Physics Research Section A: Accelerators, Spectrometers, Detectors and Associated Equipment, 380*(1–2), 228–231.
15. He, Z., Li, W., Knoll, G., Wehe, D., & Stahle, C. (2000). Measurement of material uniformity using 3-D position sensitive CdZnTe gamma-ray spectrometers. *Nuclear Instruments and Methods in Physics Research Section A: Accelerators, Spectrometers, Detectors and Associated Equipment, 441*(3), 459–467.
16. Hecht, K. (1932). Zum mechanismus des lichtelektrischen primärstromes in isolierenden kristallen. *Zeitschrift für Physik, 77*(3–4), 235–245.
17. Knoll, G. F. (2010). *Radiation detection and measurement*. Wiley.
18. Kuvvetli, I., Budtz-Jørgensen, C., Zappettini, A., Zambelli, N., Benassi, G., Kalemci, E., Caroli, E., Stephen, J. B., and Auricchio, N. (2014). A 3D CZT high resolution detector for x- and gamma-ray astronomy. In *High energy, optical, and infrared detectors for astronomy VI* (Vol. 9154, pp. 91540X). International Society for Optics and Photonics.
19. Lohstroh, A., Sellin, P., & Simon, A. (2003). High-resolution mapping of the mobility–lifetime product in CdZnTe using a nuclear microprobe. *Journal of Physics: Condensed Matter, 16*(2), S67.
20. Luke, P. (1995). Unipolar charge sensing with coplanar electrodes-application to semiconductor detectors. *IEEE Transactions on Nuclear Science, 42*(4), 207–213.
21. Owe, S. H., Kuvvetli, I., & Budtz-Jørgensen, C. (2021). Carrier lifetime and mobility characterization using the DTU 3D CZT drift strip detector. *IEEE Transactions on Nuclear Science, 68*(9), 2440–2446. https:/doi.org/10.1109/TNS.2021.3068001
22. Owe, S. H., Kuvvetli, I., & Budtz-Jørgensen, C. (2019). Evaluation of a Compton camera concept using the 3D CdZnTe drift strip detectors. *Journal of Instrumentation, 14*(1), C01020.
23. Pamelen, M. A. J. V. & Budtz-Jørgensen, C. (1998a). CdZnTe drift detector with correction for hole trapping. *Nuclear Instruments and Methods in Physics Research Section A: Accelerators, Spectrometers, Detectors and Associated Equipment, 411*(1), 197–200.
24. Pamelen, M. A. J. V. & Budtz-Jørgensen, C. (1998b). Novel electrode geometry to improve performance of CdZnTe detectors. *Nuclear Instruments and Methods in Physics Research Section A: Accelerators, Spectrometers, Detectors and Associated Equipment, 403*(2–3), 390–398.
25. Ramo, S. (1939). Currents induced by electron motion. *Proceedings of the IRE, 27*(9), 584–585.

26. Rana, V. R., III, W. R. C., Harrison, F. A., Mao, P. H., & Miyasaka, G. (2009). Development of focal plane detectors for the nuclear spectroscopic telescope array (NuSTAR) mission. In *UV, X-Ray, and Gamma-Ray Space Instrumentation for Astronomy XVI* (Vol. 7435, pp. 743503). International Society for Optics and Photonics.
27. Schlesinger, T. E. & James, R. B. (1995). Semiconductors for room temperature nuclear detector applications. In *Semiconductors and Semimetals* (p. 43).
28. Schroder, D. K. (2006). Semiconductor material and device characterization (3rd ed.). Wiley.
29. Sellin, P., Davies, A., Lohstroh, A., Ozsan, M., & Parkin, J. (2005). Drift mobility and mobility-lifetime products in CdTe: Cl grown by the travelling heater method. *IEEE Transactions on Nuclear Science, 52*(6), 3074–3078.
30. Shockley, W. (1938). Currents to conductors induced by a moving point charge. *Journal of Applied Physics, 9*(10), 635–636.
31. Suzuki, K., Seto, S., Sawada, T., & Imai, K. (2002). Carrier transport properties of HPB CdZnTe and THM CdTe: Cl. *IEEE Transactions on Nuclear Science, 49*(3), 1287–1291.
32. Sze, S. M. (2006). Semiconductor devices: Physics and technology (3rd ed.). Wiley.
33. Takahashi, T., & Watanabe, S. (2001). Recent progress in CdTe and CdZnTe detectors. *IEEE Transactions on Nuclear Science, 48*(4), 950–959.
34. Verger, L., Baffert, N., Rosaz, M., & Rustique, J. (1996). Characterization of CdZnTe and CdTe: Cl materials and their relationship to x- and γ-ray detector performance. *Nuclear Instruments and Methods in Physics Research Section A: Accelerators, Spectrometers, Detectors and Associated Equipment, 380*(1–2), 121–126.
35. Zanio, K., Akutagawa, W., & Kikuchi, R. (1968). Transient currents in semi-insulating CdTe characteristic of deep traps. *Journal of Applied Physics, 39*(6), 2818–2828.

Investigation on the Polarization Effect and Depolarization Technique of CZT Detector Under High Flux Rate X-/γ-Ray Irradiation

Xiang Chen

1   Introduction

CdZnTe (CZT) crystal belongs to ternary compound semiconductor, which has high resistivity (~10^{10} Ω·cm), wide bandgap (~1.52 eV), high γ-/X-ray stopping ability, high mobility-lifetime product, etc. [1]. Due to all these advantages, CZT detector can be operated at room temperature with a low dark current and can be simplified within centimeter-scale space. Nowadays, CZT detector has been widely applied in many fields like space exploration, nuclear medicine, and nuclear security [2–5].

However, due to the existence of high-concentration defects in CZT crystal, large amounts of γ-ray-generated carriers (electrons and holes) are trapped, forming space charges. For holes with low mobility, γ-ray-generated holes will be largely trapped to form positive space charges. The accumulation of space charges gradually enhances space charge effect (also called "polarization effect") in CZT crystal, causing the deformation of the internal electric field distribution [6–8]. Finally, the collection efficiency of γ-ray-generated carriers decreases, and the output of CZT detector occurs distortion. CZT detector shows performance failure. Under high flux rate γ-ray irradiation (higher than 10^8 γ·cm^{-2}), the polarization effect in CZT crystal is relatively common [7, 9].

Fast temporal response ability is important for high flux rate γ-ray application [10]. During the temporal response, the output current of CZT detector occurs distortion, such as the sharp peak existing at the beginning of the irradiation and the decay process from the sharp peak to the stable state [11]. This unstable process before signal stability in CZT detector is mainly attributed by the effects of deep-level defects in CZT crystal [10, 11]. Limited by the high densities of impurities

X. Chen (✉)
China Institute of Atomic Energy, Beijing, China

Northwest Institute of Nuclear Technology, Xi'an, China

and defects, excessive space charge accumulation will occur under high flux rate operation. The charged carrier dynamic process between defect occupation and charge injection from γ-ray will last several milliseconds before the signal stabilizes [10–12]. This unstable process before the stable state will greatly limit its further application.

It has been verified that sub-bandgap (infrared, abbreviated as IR) illumination can make CZT detector de-polarized [13–16]. The mechanism for the de-polarization effect by IR illumination is that IR illumination can interact directly with deep-level defects, de-trapping trapped carriers and suppressing the space charge effects in CZT crystal [13]. Plyatsko et al. tested the low-temperature photoluminescence (LTPL) spectra before and after IR illumination [14]. The test result showed that noticeable changes in the native point defects system occurred in p-type CZT crystal, and the effects of background impurities and native components were suppressed. With IR illumination, the unstable process of CZT detector in temporal response under high-intensity γ-ray irradiation would be effectively shortened and became more suitable for the optoelectronic application compared with the original crystal [14–16].

In this chapter, concentrations are mainly focused on the output current distortion of CZT detector under high flux rate γ-ray irradiation and the improvement on the output current distortion by extra IR illumination. Through the analysis on the relationship between crystal defects and detector performance distortion, this chapter can provide significant supports for the further optimization of CZT detector performance and provide high-quality CZT detector for high flux rate γ-ray irradiation measurement.

2 Defects in CZT Crystal

2.1 Source of Defects

At first, the defect source, classification, and the testing techniques are introduced. During the growth process of CZT crystal, large amounts of impurities and defects will be inevitably imported, such as from the impure elemental substance, crucible, etc., and caused by stoichiometric ratio deviation [10–12, 17–20]. These impurities and defects introduce corresponding defect levels in the energy band. On the one hand, these defect levels affect the carrier concentration in the crystal, affecting the resistivity of CZT crystal. On the other hand, these defect levels also cause scattering, trapping, de-trapping, and recombination of carriers. The complex interaction processes seriously affect the transport characteristics of carriers in CZT crystal, such as mobility-lifetime product, which in turn affect the performance of CZT detector under high flux rate γ-ray detection. In CZT crystal, it is generally recognized that shallow-level defects result in lower crystal resistivity and larger detector leakage current [12]. Deep-level defects trap carriers and form space charge in CZT crystal, resulting in the decrease of mobility and lifetime product and the deterioration in the sensitivity performance of CZT detector [10]. Therefore, a deep understanding of crystal defects is very important for the suppression of defects and performance improvement of CZT detector.

According to the source of defects in CZT crystal, these defects can be roughly divided into intrinsic defects, defects introduced by doped elements, and defects introduced by impurities.

A. *Intrinsic Defect*

In CZT crystal, intrinsic defects mainly include Cd vacancy (V_{Cd}), Te anti-site (Te_{Cd}), Te interstitial (Te_i), and defect complexes formed by these defects and impurities atoms. V_{Cd} is formed by the Cd volatilization from the raw material due to the high vapor pressure of Cd during the crystal growth under Te-rich conditions. V_{Cd} defect appears as an acceptor trap in the crystal, which greatly affect the carrier concentration and crystal resistivity [17]. Under Te-rich conditions, the excess Te can easily form secondary ionized Te anti-sites by compensating for Cd vacancies. Te_{Cd} defects manifest as deep donor levels in CZT crystal [18]. Te interstitials behave as deep principal levels in the crystal [19, 20].

B. *Defects Introduced by Doped Element*

For undoped CZT crystal, it often shows p-type and low resistivity properties. To acquire high-resistivity CZT crystal, ppm level doped elements are often used during the growth of CZT crystal, such as IIIA group elements (Al, Ga, In, etc.) and VIIA group elements (F, Cl, In, etc.) [21–25]. For example, in In-doped CZT crystal, In (IIIA group) element occupies the position of Cd and forms a shallow-level defect. This shallow defect is easy to play a self-compensation role with Cd vacancy and from defect complex, so as to realize the high resistivity of CZT crystal. VIIA group element usually occupies the position of Te element [24, 25].

C. *Defects Introduced by Impurities*

During the growth process and post-processing of CZT crystal, impurities will be inevitably introduced. The sources of impurities mainly include impurities in raw materials, impurities released from quartz crucibles at high temperatures under a long time, and impurities introduced during the process after crystal growth. Common impurities in CZT crystals are Li, Na, Mg, Al, Si, S, K, Ca, Cr, Mn, Fe, Cu, Ga, Se, etc. [26]. According to the influence on crystal properties, these impurities can be clarified as two categories. The first category is I group, III group, IV group, and VII group impurities. These impurities have a greater impact on the conductivity of CZT crystal and act as donors or acceptors. Another category is other type of impurities, which has little effects on crystal conductivity and acts as a recombination center or trap.

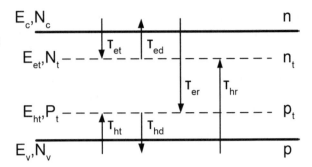

Fig. 1 Six main transit processes for n-type and p-type defects in CZT crystal [10]

2.2 Effects of Defects

When the CZT crystal is irradiated by high γ-ray flux rate, large amounts of free electron-hole pairs are generated and uniformly distributed in the whole volume. Free electrons are drifted toward the anode, and holes are drifted to the cathode. During the transport process of carriers, it will be affected by defects in CZT crystal, such as trapping, de-trapping, and recombination. It has been reported that deep levels are responsible for the polarization of CZT detector under high photon flux rates [10–12]. In the next section, the effects of deep-level defect on the performance of CZT detector is analyzed.

There are two types of defects in CZT crystal, n-type and p-type. For each type of defect, three transit processes exist, including trapping, de-trapping, and recombination. Six main transit processes can be taken to describe the effects of defects (both n-type and p-type) on carriers, as shown in Fig. 1 [10]. The lifetimes for six transit processes can be described by Eqs. (2), (3), (4), (5), and (6).

$$\tau_{et} = \frac{1}{(N_t - n_t)\,\sigma_{et}v_{th}} \tag{1}$$

$$\tau_{ed} = \frac{1}{N_t\sigma_{et}v_{th}\exp\left(-\left(E_C - E_{et}\right)/kT\right)} \tag{2}$$

$$\tau_{er} = \frac{1}{p_t\sigma_{er}v_{th}} \tag{3}$$

$$\tau_{ht} = \frac{1}{(P_t - p_t)\,\sigma_{ht}v_{th}} \tag{4}$$

$$\tau_{hd} = \frac{1}{N_V \sigma_{ht} v_{th} \exp\left(-\left(E_{ht} - E_V\right)/kT\right)} \tag{5}$$

$$\tau_{hr} = \frac{1}{n_t \sigma_{hr} v_{th}} \tag{6}$$

where N_t is the total electron trapping state density with an energy level of E_{et}, n_t is the trapped electron density at this state, P_t is the total hole trapping state density with an energy level of E_{ht}, p_t is the trapped hole density at this state, v_{th} is the thermal velocity, N_c is the available state density for the conduction band, N_v is the available state density for the valence band, σ_{et} is the electron capture cross section, σ_{er} is the recombination cross section between the free electron and the trapped holes, σ_{ht} is the hole trapping cross section, σ_{hr} is the hole recombination cross section with free electrons, k is Boltzmann's constant, and T is the Kelvin temperature.

The trapping process of carriers by defects actually reflects the process of crystal defects accumulating non-equilibrium carriers in a non-equilibrium state. The trapped carriers may also be de-trapped. The trapping and de-trapping effects of carriers by defects mainly affect the lifetime of carriers, which in turn affect the response speed of CZT detector. In CZT crystal, the trapping time of defects (10^5 s to 10^{-11} s) decreases monotonically as the defect energy level increases (0.04 eV to 0.95 eV) which the de-trapped lifetime (10^{-7} s to 1 s) increases monotonically [26, 28, 29]. This result indicates that deep-level defects can trap carriers faster than shallow-level defects and deep-level defects can de-trap carriers faster than deep-level defects. When the defect energy level is lower than 0.35 eV, the main defect types are A-center and the impurity energy level [26]. When the defect energy level is higher than 0.35 eV, the main defect types are Te anti-site and Te secondary phase-related defects [26]. At room temperature, the carrier trapping time and de-trapping time of Te secondary-phase correlated deep-level defects (with an energy level of 1.1 eV) are 12.5 ps and 38.8 s, respectively, which have a great influence on the carrier lifetime [30].

2.3 Defect Testing Technique

For semiconductor material, defect parameters mainly include defect type, energy level, concentration, cross section, etc. According to the type of defects, defects can be divided into donor-type defects and acceptor-type defects. The complex effects of defects on carrier transport process are the main reason for the deterioration of detector performance in radiation measurement. Thus, it is significant to accurately acquire the defect parameters in semiconductor material. At present, photoluminescence (PL), thermally stimulated current (TSC), and deep-level transient spectrum

(DLTS) are the commonly used methods for measuring defect parameters. In this section, we introduce the TSC testing results of CZT material, including the energy level, type, concentration, etc.

During TSC test, non-equilibrium carriers are firstly generated by light to fill the defect level as much as possible at low temperature. After the light illumination is stopped, the heating rate of CZT crystal is controlled, and the carriers captured by the defect levels are excited by heat and released and collected by the electrodes to form the detector output current signal, which is the TSC spectrum. This current signal is formed by successively emitting holes (electrons) to the valence band (conduction band) from each trap energy level from the shallow energy level to the deep energy level during the heating process. According to Eq. (7), the corresponding information such as defect energy level and concentration can be obtained from the position and area under the peak of the TSC spectrum.

$$\ln \left(\frac{T_m^4}{\beta} \right) = \frac{\Delta E_i}{k_0 T_m} - \ln \left(\frac{N_C \sigma_i v_{th} k_0}{\Delta E_i T^2} \right) \tag{7}$$

where T_m is the peak temperature. β is the heating rate, and ΔE_i is the position of i-th defect level in the forbidden band. After the acquisition of the Arrhenius curve from $\ln(T_m^2/\beta)$ and $(1/k_0 T_m)$, the i-th defect energy level ΔE_i can be obtained from the slope of the curve. The corresponding cross section σ_i can be obtained from the intercept of the curve. The concentration of the i-th defect level can be given by Eq. (8), in which V_{eff} is the effective irradiation volume of the semiconductor. G is the carrier collection efficiency. Under the assumption of uniform electric field distribution, G can be defined by Eq. (9).

$$N_{T_i} = \frac{Q_{T_i}}{2V_{eff} \times e \times G} \tag{8}$$

$$G = \frac{\mu \tau V}{L^2} \tag{9}$$

The typical defect levels of In-doped CZT crystals (from Northwestern Polytechnical University, China) are listed in Table 1, which is experimentally tested by the method of TSC technique.

Since deep levels are responsible for the polarization of CZT detector under high photon flux rates, the de-trapping lifetimes for deep defect levels on carriers

Table 1 Typical defect parameters in CZT crystal by TSC test [10]

Defect	Type	Level (eV)	Density (cm^{-3})
T1	Shallow donor	$E_C - 0.06$	6.03×10^{15}
T2	Shallow acceptor	$E_V + 0.11$	3.01×10^{15}
T3	Shallow acceptor	$E_V + 0.21$	8.12×10^{14}
T4	Deep donor	$E_C - 0.58$	5.28×10^{13}

are estimated based on the typical parameters below. The electron trapping cross section is about 6×10^{-13} cm^2, and the carrier recombination cross section between trapped electron and free hole is about 10^{-14} cm^2 [12, 26]. N_c is at the magnitude of 10^{17} cm^{-3}. N_t is at the magnitude of 10^{13} cm^{-3}. n_t is at the magnitude of 10^4 cm^{-3}. The product of k and T (300 K) is 0.026 eV. $E_c - E_{et}$ is set as 0.58 eV. The thermal velocity is about 10^7 cm/s. Based on these parameters, the estimated de-trapping lifetimes for T1, T2, T3, and T4 defects are at the magnitude of 0.1 ns, 1 ns, 100 ns, and 1 s, respectively. The estimated de-trapping lifetime of the deep level is in accordance with the results in [26], which is also at the magnitude of second. These estimated lifetimes are important for the understanding of the unstable output signal of CZT detector under high flux rate γ-ray radiation.

3 Temporal Response Experiment

3.1 Sensitivity

Sensitivity $S(E)$ is one of the key parameters for CZT detector in the application of high flux rate radiation measurement, reflecting the ability to accurately obtain the flux rate of γ-ray. This parameter is defined as the ratio between the output current of CZT detector I and the flux rate of γ-ray $\phi(E)$ incident vertically from the front end of CZT detector. The unit for sensitivity is A·γ^{-1}·cm^{-2}·s^{-1} (or C·γ^{-1}·cm^{-2}·s^{-1}). Steady-state radioactive sources such as ^{60}Co and ^{137}Cs are often used for the sensitivity test of CZT detector.

$$S(E) = \frac{I}{\phi(E)} \qquad (10)$$

3.2 Sensitivity Test

The temporal response of CZT detector to γ-ray is calibrated on a ^{60}Co γ-ray radiation instrument with an activity of about 2600 Ci. The output current of CZT detector is recorded by a micro electrometer, which can acquire current signals smaller than pA. The working bias voltage (600 V) is supplied by a high-voltage source. The layout of the system is shown in Fig. 2. The dose rate at the position of CZT detector was measured by an ionization chamber.

Since the activity of ^{60}Co source is high to about 2600 Ci, this radioactive source is surrounded by thick lead walls. A small hole is set in the wall to utilize γ-ray. A lead shutter is placed at the front of the hole. The state conversion between radiation and non-radiation is achieved by lifting or putting down the shutter. To obtain the duration time of the lifting process, a fast response photoelectric tube (named GD) is used. GD has a fast time response of ns [31]. The testing result is shown in Fig.

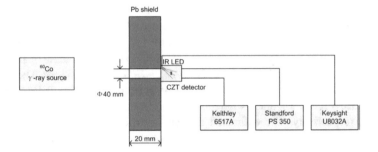

Fig. 2 Layout of temporal response system [10]

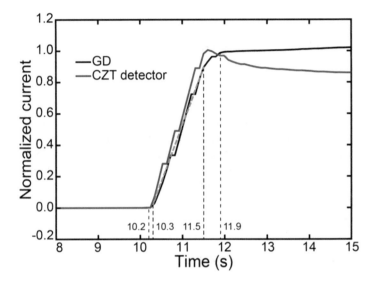

Fig. 3 Testing results on the duration time for the lifting process of the shutter [10]

3, where the black line is the response curve of GD and the red line is the response curve of CZT detector. Thus, about 1.7 s is needed for the lifting process of the shutter. This result verifies that the linear increases of the current (both from the GD and CZT detector) are related to the lifting process of the shutter and the sharp peak is only related to the CZT detector.

When CZT detector is set at 6 m away from the cobalt source device, the measured flux rate at the front end of CZT detector is 3.903 R/min. According to the conversion relationship between flux rate and γ-ray intensity given by Eq. (11), the incident γ-ray intensity at the measuring point is 1.07×10^8 γ·cm^{-2}·s^{-1} [31]. Under different working voltages, the output signal of CZT detector is shown in Fig. 4.

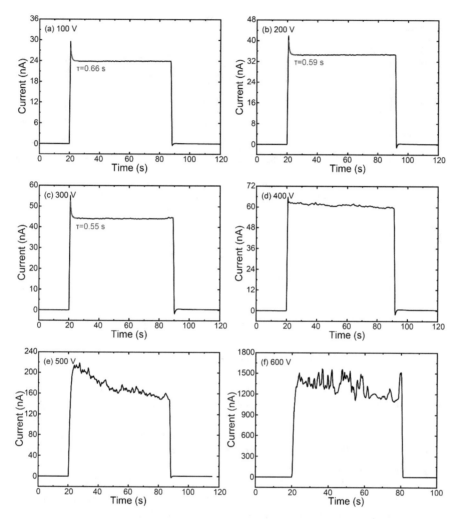

Fig. 4 Output signals of 1#CZT detector change with the different working voltages under the flux rate of 3.903 R/min

$$\frac{d\phi}{dt}\left(MeV \cdot cm^{-2} \cdot s^{-1}\right) = I\left(R \cdot min^{-1} \times 3.405 \times 10^{7}\right) \qquad (11)$$

As shown in Fig. 4d–f, when the working voltage is higher than 300 V, the output signal of 1#CZT detector is distorted. This result indicates that under the high working voltage condition, the output current of 1#CZT detector is unstable, and the test results cannot be used for high flux rate γ-ray measurement. When the working voltage is not higher than 300 V, the output current of 1#CZT detector has an overshoot peak before reaching the stable output state (flat value), and the

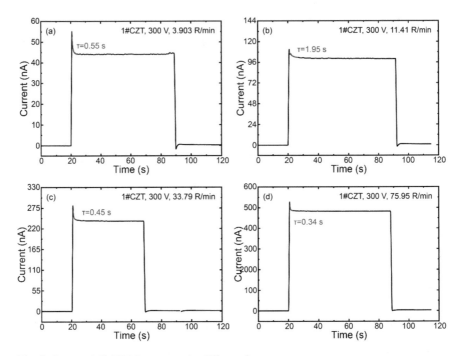

Fig. 5 Output of 1# CZT detectors under different flux rates

descending process of the overshoot peak follows the mono-exponential decay law. The attenuation constant decreases gradually as the operating voltage increases, as shown in Fig. 4a–c. Considering the operating working voltage is higher, the decay constant of the trailing edge of the overshoot peak becomes smaller, and the detector output signal can reach a stable value in a shorter time.

Steady-state flux rate response refers to the change law of CZT detector sensitivity with radiation flux rate. Due to the high concentration of defects in CZT crystal, the effect of defects on the γ-ray-generated carriers results in the loss of carriers, which reduces the carrier collection efficiency and makes the detector steady-state flux rate response nonlinear. When the radiation flux rates at the front end of the detector are 3.903 R/min, 11.41 R/min, 33.79 R/min, and 75.95 R/min, respectively, the output signals of the 1#CZT detector are shown in Fig. 5.

Under the conditions of different radiation flux rates, the output current of the 1#CZT detector has an overshoot peak (Fig. 5) before reaching the stable output state (flat value), and the descending process of the overshoot peak follows the mono-exponential decay law. On the whole, the attenuation constant obtained by fitting the output signal of the 1#CZT detector gradually decreases with the increase of the radiation flux rate.

According to the definition of steady-state sensitivity, the plateau current is used as the output current of the CZT detector when it is working stably, and the relationship between the sensitivity of 1#CZT detector and the flux rate is obtained.

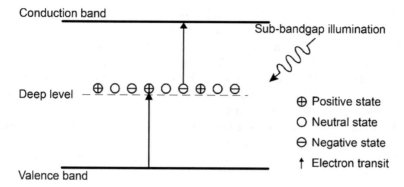

Fig. 6 Schematics on the generation of free carriers by sub-bandgap illumination

The results show that the sensitivity of the CZT detector has a higher nonlinearity with the change of the radiation flux rate. 3.32×10^{-16} C·cm^2/MeV decreased to 1.87×10^{-16} C·cm^2/MeV, decreasing 43.8%.

3.3 Sub-bandgap Illumination Optimization

Sub-bandgap light refers to light with a wavelength longer than the wavelength corresponding to the forbidden band width of CZT crystal. The bandgap of CZT crystal is about 1.52 eV, corresponding to the wavelength of light about 817 nm. Thus, sub-bandgap light is also called as near-infrared light (infrared, IR).

Based on the Lambert-Beer law, when the energy of the incident visible light (referred to as incident light) is higher than the band gap of CZT crystal, most energy of the incident light will be deposited near the crystal incidence surface (~1 μm), generating electron-hole pairs by ionization. The energy of sub-bandgap light is lower than the forbidden band width of the crystal, which will directly interact with the deep-level traps of CZT crystal, causing the trapped carriers to be released, which is equivalent to increasing the carrier mobility lifetime product, as shown in Fig. 6 [13].

It should be noted that sub-bandgap light can also ionize defects that are originally in a neutral state, causing electrons to transit from the defect energy level to the conduction band [13]. The two action modes of sub-gap light on deep-level defects are in a competitive relationship, which should be analyzed and determined according to the actual situation. Under the condition of sub-gap-band illumination, the increase of surface current and bulk current of CZT crystal is related to the de-trapping effect of trapped carriers on crystal defects. The influence of suitable sub-gap-band illumination on the distribution of electric field intensity in the crystal is more obvious [13, 32].

During the experiment, IR illumination is always irradiated on the surface of 2#CZT crystal (5 mm × 5 mm × 2 mm) before and after γ-ray, as shown in Fig. 2. IR photocurrent refers to the output current of CZT detector generated by IR illumination. With the change of IR illumination intensity, the IR photocurrents of CZT detector are different, as shown in Fig. 7. I_{base} means the sum between the intrinsic dark current (2.8 nA) and IR photocurrent of 2#CZT detector before γ-ray irradiation. I_{signal} is the output current of 2#CZT detector irradiated both by γ-ray and IR illumination. The difference between I_{signal} and I_{base} refers to the net current induced by γ-ray, which is also called as the stable state current.

As shown in Fig. 7a, the stable state current of 2#CZT detector is about 23.5 nA under different IR photocurrents. For the analysis of the shape changes, the IR photocurrent of 2#CZT detector is deducted from I_{signal}. However, IR photocurrent is also used as the key parameter to describe the effect of IR illumination on the performance of 2#CZT detector. This is because that the effect of IR illumination is transiting electrons from the valence band to the defect level to improve the defect occupation. After the deduction of IR photocurrent, the temporal responses of 2#CZT detector to high flux rate γ-ray under different IR photocurrents are aligned at the same I_{base}. The effects of IR photocurrent on the sharp peaks are shown in Fig. 7b. The flux rate at the position of 2#CZT detector is measured as 5.403 Gy/h. As shown in Fig. 7b, while no IR illumination is operated on 2#CZT detector, a sharp peak appears at the beginning of γ-ray irradiation. Following the sharp peak, a stable state is achieved after a decaying process. The time scale of the sharp peak is at the magnitude of second. Since the switch-on time of radiation and RC constant of the circuit are both shorter than 10 ms, this sharp peak and the decay process are theoretically not related to the circuit factors. With the increase of IR photocurrent, the current amplitude of the sharp peak decreases and nearly disappears when the IR photocurrent reaches to larger than 40 nA. The stable state current of 2#CZT detector increases slightly with the increase of IR photocurrents. Under different IR photocurrents, the temporal responses of 2#CZT detector to γ-ray radiation are carefully analyzed, including shape changes (such as the sharp peak, the stable current value, and the decay process), current uncertainty, and flux rate response.

4 Analysis and Discussion

4.1 Shape Changes with no IR Illumination

With no IR illumination and γ-ray radiation, the output currents of 1#CZT and 2#CZT detectors are attributed by the stable dynamic process between carrier generation (such as carrier de-trapping and electrode injection) and carrier disappearance (trapping, recombination, and so on). When γ-ray is incident, the stable dynamic process is broken, and a new stable state will be built under γ-ray irradiation. As shown in Fig. 7b, the built-up process of the stable state consists of the linear

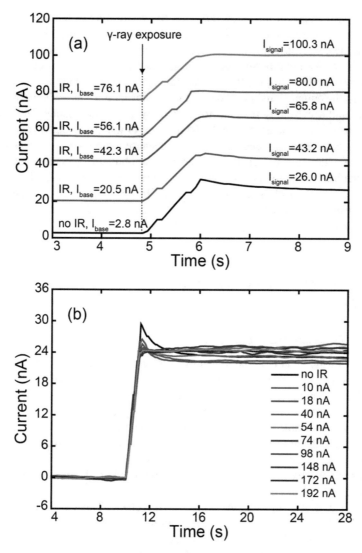

Fig. 7 Temporal responses of 2#CZT detector to γ-ray under different IR photocurrents. (**a**) Effects of IR photocurrent on I_{base}. (**b**) Effects of IR photocurrent on sharp peak, and I_{base} values are deleted [10]

increase of current, a sharp peak existing at the beginning of γ-ray radiation, a stable state, and a decay process from the sharp peak to the stable state. The sharp peak might be due to the excessive carrier traps at the beginning of γ-ray irradiation, which would de-trap the redundant carriers before the stable state. The decay process is the built-up of a new charged carrier dynamic equilibrium between the defect occupation and the charge injection from γ-ray irradiation.

With the increase of the γ-ray flux rate, the decay time of the trailing edge of the overshoot peak is gradually shortened, and the working voltage had little effect on the decay constant. These two features indirectly reflect that the carrier transport process is not the main reason for the overshoot peak, and the carrier de-trapping process by deep-level traps is the intrinsic factor for the overshoot peak.

Ideally, the output signal plateau of CZT detector should change linearly with the flux rate, and accordingly the detector sensitivity should not change with the flux rate. However, in the actual test, with the increase of the flux rate, the sensitivity of 2#CZT detector gradually decreases, mainly because more γ-ray-generated carriers recombine. The main reason for the carrier recombination process might be the recombination caused by the recombination center near the Fermi level and caused by the accumulation of space charges caused by the polarization of the electric field intensity distribution. Further studies will be operated to analyze which one dominates the recombination process in CZT crystal.

4.2 Effects of IR Illumination on the Temporal Response of CZT Detector

A. Sharp Peaks and Stable Current Values

IR illumination can help the transit of electron from valence to the defect level. This transit process can be taken as a prefilling process for the defect level, which will reduce the number of trapping carriers generated by γ-rays. As shown in Fig. 8, with the increase of IR photocurrent, the sharp peak values of 2#CZT detector decrease and its stable current values increase. The decrease of sharp peak value represents that the excessive carrier traps are suppressed. The increase of the stable current values is attributed by the effects of the prefilling process. The 50-nA photocurrent means that about 3.1×10^{11} free carriers (holes or electrons) are generated within 1 s. The corresponding generated free carrier concentration in the 5 mm $\times$ 5 mm $\times$ 2 mm volume is 6.2×10^{12} cm^{-3}. This carrier concentration is comparable to the T4 defect concentration in Table 1, which is reported as 5.82×10^{13} cm^{-3} by TSC technique. When IR photocurrent reaches to 50 nA, the sharp peak in the output current of 2#CZT detector disappears, and the stable current value keeps constant. The stable current value is obviously improved from 23.2 nA (IR photocurrent = 0 nA) to 25.1 nA (IR photocurrent = 50 nA). This increased stable current value refers to the prefilling effects on the carrier trapping process. Thus, more carriers generated by γ-ray are collected, and the collection efficiency of the generated carriers in the 2#CZT detector is improved by 8.2%. Before the IR photocurrent reaches to 150 nA (see in Fig. 7), the stable current value of CZT detector keeps constant and decreases with the further increase of IR photocurrent. This decrease of the stable current value might be related to the stronger depletion layer near the cathode caused by the excess carrier injection. To keep the stability

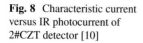

Fig. 8 Characteristic current versus IR photocurrent of 2#CZT detector [10]

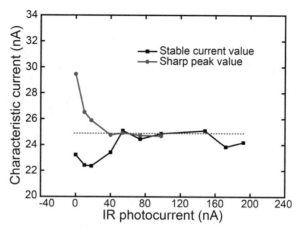

of CZT γ-ray detector, the IR photocurrent of CZT γ-ray detector should be limited to 50–150 nA.

B. *Decay Process*

With no IR illumination, a sharp peak appears at the beginning of the irradiation, as shown in Fig. 7b. With the increase of IR photocurrent, the sharp peak disappears gradually. Deep levels are reported to be responsible for the polarization of CZT detector under high photon fluxes [27]. IR illumination has a "prefilling" effect on the deep level. This sharp peak is probably related to the deep defect level. Assuming that the de-trapping process follows exponential decay order, a bi-exponential fitting method is used to analyze the decay process, as shown in Fig. 9 [10]. The fitting parameters are listed in Table 2.

When IR photocurrent is over 18 nA, the bi-exponential fitting time constants become to the same value within the time of the red lines in Fig. 9. Single exponential fitting method is used, and the fitting results are listed in Table 2. With the increase of the IR photocurrent, τ_1 gets smaller and finally disappears when the IR photocurrent is over 18 nA, as shown in Fig. 9. τ_2 also gets smaller from 1.85 to 0.47 s, while the IR photocurrent increases from 0 to 40 nA. This result means that the influence of deep defect level on the temporal response of CZT γ-ray detector is at the magnitude of second, which is in good accordance with the estimated de-trapping lifetime of deep defect level. The deep level dominates the unstable dynamic process in the temporal response of CZT detector. For the experimental CZT crystal, the main deep donor defect exists at the level of E_C-0.58 eV, related to the Te antisite-related deep donors. IR illumination can actually impress the effect of the defect occupation and shorten the unstable dynamic process to a stable state under high flux rate γ-ray irradiation.

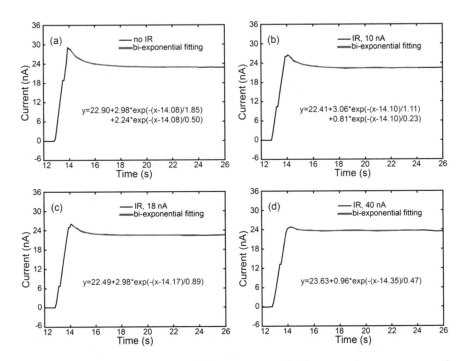

Fig. 9 Bi-exponential fitting curves of the decay process in the temporal response curves of CZT detector under different IR photocurrents. (**a**) IR photocurrent = 0 nA. (**b**) IR photocurrent = 10 nA. (**c**) IR photocurrent = 18 nA. (**d**) IR photocurrent = 40 nA [10]

Table 2 Bi-exponential fitting time constants of CZT detector under different IR photocurrents [10]

IR photocurrent (nA)	Bi-exponential fitting time constants (s)	
	τ1	τ2
0	0.50	1.85
10	0.23	1.11
18	–	0.89
40	–	0.47

C. Current Uncertainty Analysis

The net current uncertainty of 2#CZT γ-ray detector induced by different IR photocurrents is analyzed in Fig. 10 [10].

With no IR illumination, the current uncertainty is smaller than 0.2%. With the increase of IR photocurrent, the current uncertainty also increases. When the IR photocurrent is within 50 nA–150 nA, the current uncertainty keeps a stable value of about 0.84%. It is understandable that the increase of the current uncertainty is related to the IR photocurrent. When the IR illumination is turned off, the net current of CZT detector is about 24 nA. Compared with this net current, IR photocurrent

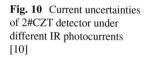

Fig. 10 Current uncertainties of 2#CZT detector under different IR photocurrents [10]

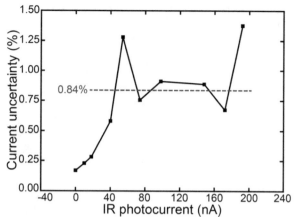

is much larger. The current uncertainty generated by IR photocurrent cannot be ignored for the net current of CZT detector with no IR illumination. This is the reason for the increase of current uncertainty under IR illumination. When the IR photocurrent of CZT detector is smaller than 200 nA, the current uncertainty is lower than 1.4%. Under this condition, the influence of current uncertainty induced by IR illumination can be ignored in the temporal response of CZT detector.

D. Flux Rate Response

Flux rate response of 2#CZT detector under different γ-ray intensities is also investigated. Near the hole in the lead wall, γ-ray irradiation does not distribute uniformly. To assure the uniform γ-ray irradiation on CZT detector, the CZT detector is set at the place of 2.2 m and 3.1 m away from ^{60}Co source. The flux rate at the position of 2.2 m and 3.1 m is measured by UNIDOS, which are 5.403 and 2.590 Gy/h, respectively. Sensitivity versus IR photocurrent curves of 2#CZT detector are shown in Fig. 11.

When no IR is illuminated on CZT detector, the absolute sensitivity difference is 9.7%. Under different IR photocurrents (90 nA, 100 nA, and 110 nA), the absolute response differences are 5.2%, 3.9%, and 2.6%, respectively, much lower than that with no IR illumination.

This difference is mainly related to the carrier combination under different γ-ray intensities. Based on Eq. (3), the lifetime of electron recombination τ_{er} is reciprocal to the trapped hole density p_t. With the increase of γ-ray intensity, p_t also increases. The increase of p_t results in the decrease of τ_{er}, meaning that the probability of electron recombination becomes higher. More electrons will be trapped and recombined by the deep level. With the increase of γ-ray intensity, the sensitivity of CZT detector becomes lower, and this is the reason for the sensitivity difference of 9.7% with no IR illumination under 5.403 Gy/h and 2.590 Gy/h. IR illumination can prefill the deep level and suppress the effects of deep levels on the sensitivity of CZT γ-ray detector.

Fig. 11 Flux rate responses of CZT detector under 5.403 and 2.590 Gy/h [10]

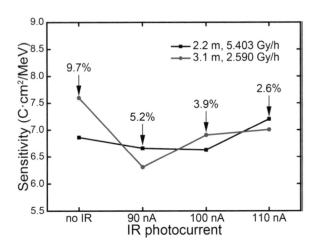

With the increase of IR illumination, the sensitivity difference of CZT detector becomes lower. The flux rate response of CZT detector can be improved by IR illumination.

5   Conclusion

In this chapter, the polarization effect and de-polarization technique of CZT detector under high flux rate X-ray irradiation are analyzed, including the composition of the measurement system, the influence of polarization on the performance of the detector, and the IR illumination depolarization technology for the polarization effect of the CZT detector. The results show that:

1. Polarization effects on the temporal response of CZT detector

During the temporal response of CZT detector, the output currents occur many distortions, such as the sharp peak at the beginning of γ-ray radiation, the nonlinearity relationship between the stable output current and the flux rate, and the unstable current under high working voltages. These phenomena are all greatly related to the effects of the deep-level defects in CZT crystal.

2. Depolarization effects of IR illumination on CZT detector

Under high flux rate γ-ray irradiation, IR illumination is experimentally verified to be effective to suppress the unstable process (including sharp peak and stable current nonlinearity with the flux rate). Reasons for this optimization are attributed to the IR prefilling effect on the deep level, which can suppress the effects of deep-level defects on γ-ray-generated carriers. IR illumination can improve the stability of CZT detector under γ-ray irradiation. With the increase of IR photocurrent, the sharp peak disappears gradually, and the temporal response of CZT detector follows the intensity of γ-ray irradiation without transition state. The fitting time constant of the decay process is at the magnitude of second, which is in good accordance with the estimated de-trapping lifetime of deep level. The stable current of CZT detector to γ-ray irradiation can be largely improved due to the prefilling effects of IR illumination on the defect level.

This chapter investigates the polarization effect and de-polarization technique of CZT detector under high flux rate γ-ray radiation. It can solve the instability problems during the sensitivity calibration process of CZT γ-ray detector, which will provide significant experimental supports for the further application of CZT material.

References

1. Zhu, R. Y. (2018). The next generation of crystal detector. *Radiation Detection Technology and Methods, 9593*, 1–12.
2. Henric, S. K., Danie, S., Fiona, A. H., et al. (2016). X-ray polarimetry with the Polarization Spectroscopic Telescope Array (PolSTAR). *Astroparticle Physics, 75*, 8–28.
3. Jaesub, H., Branden, A., Jonathan, G., et al. (2013). Tiled array of pixelated CZT imaging detectors for ProtoEXIST2 and MIRAX-HXI. *IEEE Transactions on Nuclear Science, 60*(6), 4610–4617.
4. Lane, W. D., Lori, B. C., Tapan, G., et al. (2010). Reduced isotope dose with rapid SPECT MPI imaging: Initial experience with a CZT SPECT camera. *Journal of Nuclear Cardiology, 17*(6), 1009–1014.
5. Zha, G. Q., Yang, J., Xu, L. Y., et al. (2014). The effects of deep level traps on the electrical properties of semi-insulating CdZnTe. *Journal of Applied Physics, 115*(4), 1–4.
6. Bolotnikov, A. E., Camarda, G. S., Cui, Y., et al. (2009). Internal electric-field-lines distribution in CdZnTe detectors measured using X-ray mapping. *IEEE Transactions on Nuclear Science, 56*(3), 791–794.
7. Camarda, G. S., Bolotnikov, A. E., Cui, Y. G., et al. (2008). Polarization studies of CdZnTe detectors using synchrotron X-ray radiation. *IEEE Transactions on Nuclear Science, 55*(6), 3725–3730.
8. Guo, Y., Zha, G. Q., Li, Y. R., et al. (2019). Effect of transient space-charge perturbation on carrier transport in high-resistance CdZnTe semiconductor. *Chinese Physics B, 28*(11), 1–10.
9. Wang, X., Xiao, S. L., Li, M., et al. (2013). Further polarization effect of CdZnTe detectors under high flux X-ray irradiation. *High Power Laser and Particle Beams, 25*(3), 773–777.
10. Chen, X., Song, Z. H., Lu, Y., et al. (2021). Improvement on the temporal response of CZT γ-ray detector by infrared illumination. *IEEE Transactions on Nuclear Science, 68*(10), 2533–2538.
11. Prokesch, M., Bale, D. S., & Szeles, C. (2010). Fast high-flux response of CdZnTe X-ray detectors by optical manipulation of deep level defect occupations. *IEEE Transactions on Nuclear Science, 57*(4), 2397–2399.
12. Du, Y. F., Lebanc, J., Possin, G. E., et al. (2012). Temporal response of CZT detectors under intense irradiation. *IEEE Transactions on Nuclear Science, 50*(4), 480–484.
13. Teague, L. C., Washington, A. L., II, Duff, M. C., et al. (2012). Photo-induced currents in CdZnTe crystals as a function of illumination wavelength. *Journal of Physics D: Applied Physics, 45*(10), 1–6.
14. Plyatsko, S. V., Petrenko, T. L., & Sizov, F. F. (2009). Defects instability under IR laser treatment in p-CdZnTe crystals. *Infrared Physics & Technology, 52*, 57–61.

15. Dedic, V., Franc, J., Sellin, P. J., et al. (2012). Study on electric field in Au/CdZnTe/In detectors under high fluxes of X-ray and laser irradiation. *Journal of Instrumentation, 7*(2), 1–10.
16. Suzuki, K., Mishima, Y., Masuda, T., et al. (2020). Simulation of the transient current of radiation detector materials using the constrained profile interpolation method. *Nuclear Instruments and Methods in Physics Research A, 971*, 164128.
17. Li, G., Sang, W. B., Min, J. H., et al. (2008). Study on the defect energy levels of high resistivity In-doped CdZnTe crystals. *Journal of Inorganic Materials, 23*(5), 1049–1053.
18. Wang, P. F., Nan, R. H., & Jian, Z. Y. (2017). The effects of deep-level defects on the electrical properties of Cd0.9Zn0.1Te crystals. *Journal of Semiconductors, 38*(6), 1–6.
19. Guo, R. R., Jie, W. Q., Wang, N., et al. (2015). Influence of deep level defects on carrier lifetime in CdZnTe:In. *Journal of Applied Physics, 117*(9), 1–5.
20. Wei, S. H., & Zhang, S. B. (2002). Chemical trends of defect formation and doping limit in II-VI semiconductors: The case of CdTe. *Physical Review B, 66*(15), 1–10.
21. Fiederle, M., Fauler, A., Konrath, J., et al. (2004). Comparison of undoped and doped high resistivity CdTe and (Cd,Zn)Te detector crystals. *IEEE Transactions on Nuclear Science, 51*(4), 1864–1868.
22. Wang, T., Ai, X., Yin, Z., et al. (2019). Study on a co-doped CdZnTe crystal containing Yb and In. *Crystal Engineering Communication, 21*(16), 2620–2625.
23. Yang, F., Jie, W. Q., Zha, G. Q., et al. (2020). The effect of indium doping on deep level defects and electrical properties of CdZnTe. *Journal of Electronic Materials, 49*(2), 1243–1248.
24. Won, J. H., Kim, K. H., Cho, S. H., et al. (2008). Electrical properties and X-ray spectrum of semi-insulating CdZnTe:Pb crystals. *Nuclear Instruments & Methods in Physics Research A, 586*, 211–214.
25. Roy, U. N., Weiler, S., Stein, J., et al. (2011). A. Internal electric field estimation, charge transport and detector performance of as-grown Cd0.9Zn0.1Te: In by THM. *IEEE Transactions on Nuclear Science, 58*(4), 1949–1952.
26. Zhang, J. X., Liang, X. Y., Min, J. H., et al. (2019). Effect of point defects trapping characteristics on mobility-lifetime ($\mu\tau$) product in CdZnTe crystals. *Journal of Crystal Growth, 519*, 41–45.
27. Rejhon, M., Franc, J., Dedic, V., et al. (2016). Analysis of trapping and de-trapping in CdZnTe detectors by Pockels effect. *Journal of Physics D: Applied Physics, 49*, 1–8.
28. Luo, X. X., Zha, G. Q., Xu, L. Y., et al. (2019). Improvement to the carrier transport properties of CdZnTe detector using sub-band-gap light radiation. *Sensors, 19*, 1–10.
29. Gaubas, E., Ceponis, T., Deveikis, L., et al. (2020). Study of the electrical characteristics of CdZnTe Schottky diodes. *Materials Science in Semiconductor Processing, 105*, 1–11.
30. Suh, J., Hong, J., Franc, J., et al. (2016). Tellurium secondary-phase defects in CdZnTe and their association with the 1.1-eV deep trap. *IEEE Transactions on Nuclear Science., 63*(5), 2657–2661.
31. Chen, X., Zhang, Z. C., Zhang, K., et al. (2020). Study on the time response of a barium fluoride scintillation detector for fast pulse radiation detection. *IEEE Transactions on Nuclear Science, 67*(8), 1893–1898.
32. Zhang, K., Hu, H. S., Song, Z. H., et al. (2021). Experimental investigation on pulsed gamma-ray fluence rate effect on Yb-doped yttrium aluminum garnet scintillator. *Review of Scientific Instruments, 92*(6), 063304.

Research on the Performance of CZT Detector in Alpha Particle Detection

Yu Xiang and Long Wei

1 Introduction

α particle detection is widely applied in many fields, such as determination of radioactive materials, radiation protection, environment protection, nuclear fuel facility monitor, and radionuclide therapy [1, 2]. Since α particle is heavy charged, it loses energy primarily through the excitation and ionization of target atom. According to this principle, many kinds of detectors are invented. They can be categorized as gaseous detectors, scintillator detectors, and semiconductor detectors. In gaseous detector, after the collision with α particle, the orbital electron gets rid of the bound with gas atom and becomes a free electron. An ion is also created while the electron escapes. The transport of the electron-ion pair in the electric field produces change of induced charge on the electrodes, which is collected by the connected electronics.

In scintillator detector, the generation process of the signal representing energy deposition of alpha particle is quite different and complicated. Firstly, a scintillator atom is excited by the incident alpha particle. Secondly, fluorescence generates

Y. Xiang
Beijing Engineering Research Center of Radiographic Techniques and Equipment, Institute of High Energy Physics, Chinese Academy of Sciences, Beijing, China

School of Nuclear Science and Technology, University of Chinese Academy of Sciences, Beijing, China

L. Wei (✉)
Beijing Engineering Research Center of Radiographic Techniques and Equipment, Institute of High Energy Physics, Chinese Academy of Sciences, Beijing, China

School of Nuclear Science and Technology, University of Chinese Academy of Sciences, Beijing, China

Jinan Laboratory of Applied Nuclear Science, Jinan, China
e-mail: weil@ihep.ac.cn

L. Abbene, K. (Kris) Iniewski (eds.), *High-Z Materials for X-ray Detection*,
https://doi.org/10.1007/978-3-031-20955-0_7

from the deexcitation of the atom. Then the fluorescence is transformed into photoelectrons on the photocathode of photomultiplier. Finally, the photoelectrons rapidly propagate in photomultiplier and collected by the anode.

In semiconductor detector, the signal's generation is almost the same as the one in gaseous detector. The only difference between them is that in semiconductor, electron-hole pairs rather than electron-ion pairs are created. While an electron becomes free and moves to the conduction band, there is an empty state left in the valence band. If an electron comes from another state to fill the empty state, the current generated in this process is the same as the one from the drift of a particle with one positive electric charge from the empty state to the state that the electron used to occupy. Hence, to efficiently describe the current in valence band, a virtual particle named hole is introduced to represent the empty state. The movement of hole is achieved by electron.

Due to high atomic number and density, semiconductor has greater stopping power than gas. Thus, the range of α particle is usually only a few tens of microns in semiconductor, while in gas the value is three orders of magnitude higher, and that's why the volume of alpha particle semiconductor detector is permitted far smaller than gaseous. Besides, in general, the ionization energy of semiconductor is rather low, only a few eVs, meaning that numbers of e–h pairs can be generated by one interaction, and the signal fluctuation is narrow.

The benefits of semiconductor shown above confirm that semiconductor is a good choice for α particle detection. In present, the most popular semiconductor materials in nuclear radiation detection are HPGe (high-purity germanium) and Si. HPGe can achieve an extremely fine energy resolution better than 21 keV FWHM at 5486 keV [3]. However, HPGe can't work normally unless the environment temperature under a few Ks. Si also have an astonishing energy resolution of 16 keV FWHM at 5486 keV [4]. Perovskite $CsPbBr_3$ is a member of new-generation semiconductor materials, able to be operated in room temperature, but has an unsatisfying resolution of 15% at 5486 keV [5]. GaN holds good resistance to high temperature, able to be applied under extreme conditions. Its energy resolution is about 2.2% at 5486 keV [6]. CZT (CdZnTe, cadmium zinc telluride) is a second-generation semiconductor material. In the past three decades, it has attracted the attention of many researchers because of its strong points comparing with other semiconductor materials. The physical properties for CZT and some other semiconductor materials are shown in Table 1. CZT has a wide forbidden band and high resistivity, which keeps the leaking current of CZT detector at a low level and makes it able to work normally at room temperature. The atomic number of CZT is large, resulting in strong stopping power for X-rays, γ-rays, and charged particles. The energy resolution of CZT for α particle is reported as 1.18%@5486 keV, which is better than most of other semiconductor detectors [7].

CZT is very popular in X-ray and gamma-ray detection. It can be manufactured as Compton camera and SPECT or PET system, playing an indispensable role in cosmic radiation detection, medical diagnosis, and homeland security. However, the research on CZT's performance in alpha particle detection is really rare, in spite of its good potential in this field. In this chapter, an α particle detecting system and

Table 1 Physical properties for different semiconductor materials [8]

Material	Atomic number	Density (g/cm³)	Forbidden band width (eV)	Ionization energy (eV)	Resistivity (Ω·cm)	$\mu_e \tau_e$ [a] (cm²/V)	$\mu_h \tau_h$ [b] (cm²/V)
Si	14	2.33	1.12	3.62	$>10^4$	>1	>1
Ge	32	5.33	0.67	2.96	50	~1	>1
GaAs	31,33	5.32	1.43	4.2	10^7	8×10^{-5}	4×10^{-6}
GaAs	31,33	5.32	1.43	4.2	10^7	10^{-5}	10^{-6}
InP	15,49	4.78	1.35	4.2	10^6	5×10^{-6}	$<2 \times 10^{-5}$
CdTe	48,52	6.20	1.44	4.43	10^9	10^{-3}	10^{-4}
$Cd_{0.9}Zn_{0.1}Te$	48,30,52	5.78	1.57	4.64	$10^{10} \sim 10^{11}$	$10^{-3} \sim 10^{-2}$	$10^{-6} \sim 10^{-5}$
$Cd_{0.8}Zn_{0.2}Te$	48,30,52	6.02	$1.5 \sim 2.2$	5.0	$10^{10} \sim 10^{11}$	10^{-3}	$10^{-6} \sim 10^{-5}$
HgI_2	80,53	6.40	2.13	4.2	10^{13}	10^{-4}	$10^{-5} \sim 10^{-4}$
PbI_2	82,53	6.20	$2.3 \sim 2.6$	4.9	10^{12}	10^{-6}	10^{-7}
TlBr	81,35	7.56	2.68	6.5	10^{12}	10^{-5}	10^{-6}

[a] Mobility-lifetime product of electron
[b] Mobility-lifetime product of hole

a detector depending on CZT is created. A radiation source made up of Am-241 and Pu-239 is used to test the performance of the system. An energy resolution of 1.47% FWHM at 5.486 MeV and 1.32% at 5.157 MeV has been achieved in vacuum (0.01 atm). Simultaneously, the simulations which concern about the transport of charge carriers and the interaction between α particles and CZT has been completed. To approve the reasonableness of the simulations, a simulated spectrum is compared with the one from experiment, and they agree with each other. In this process, a simulation methodology is developed and growing mature.

2 CZT Detecting System

The CZT α particle detecting system can be taken apart in three main modules, a CZT detector, an electronics, and a DAQ program. The scheme of the detection system is shown in Fig. 1. In this part, the function of each module in detection will be introduced.

2.1 CZT Detector

In CZT, the signal comes from the change of induced charge on electrodes and is proportional to the energy deposition of incident alpha particle. In planar CZT detector, the change of induced charge collected by the electronics, marked as Q_i, can be expressed by Hecht formula [9].

$$Q_i = Q_e + Q_h$$
$$= \text{Ne}\frac{\mu_e\tau_e E}{d}\left(1 - \exp\left(-\frac{z}{\mu_e\tau_e E}\right)\right) + \text{Ne}\frac{\mu_h\tau_h E}{d}\left(1 - \exp\left(-\frac{d-z}{\mu_h\tau_h E}\right)\right) \tag{1}$$

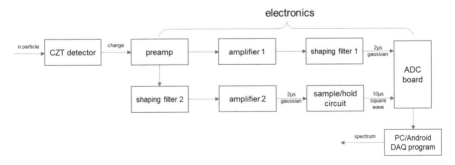

Fig. 1 Scheme of the detection system

where Q_e is the induced charge caused by electrons, Q_h is the induced charge caused by holes, N represents the number of electron-hole pairs in the detector, e is the elementary charge, E is electric field, d is the thickness of the detector, and z is the distance from the position where the electron-hole pair generates to detector's anode. As a result of the limitation of CZT crystal growth method, the detector is usually a few millimeters or centimeters thick. Therefore, the values of $\frac{z}{\mu_e \tau_e E}$ and $\frac{d-z}{\mu_h \tau_h E}$ are nearly 0, bringing out a simplified formula of Eq. (1) as

$$Q_i = Ne \left(1 - \frac{1}{2dE} \left(\frac{z^2}{\mu_e \tau_e} + \frac{(d-z)^2}{\mu_h \tau_h} \right) \right) \qquad (2)$$

It is obvious that the value of induced charge relies on the mobility-lifetime products of hole and electron. Since the hole mobility-lifetime product of CZT is about two orders of magnitude lower than electron's, the trapping of holes causes a wide fluctuation of induced charge, deeply affecting the energy resolution of CZT detector. It is reported that in the γ-ray spectra acquired by a planar CZT detector, the 662 keV peak is unrecognized, while the energy resolution at 59.6 keV is about 7% FWHM [10].

To mitigate the influence of hole trapping, single polarity charge sensing technique is developed, named after its function. The anode weighting potential in a detector manufactured basing on this technique is around 0 in almost the whole bulk, excluding the area surrounding the anode, where the weighting potential increases rapidly to 1. For a deeper understanding of the technique's benefit, here the meaning of weighting potential will be described. In the end of the 1930s, Shockley and Ramo raised a theorem to calculate induced charge generated by a moving charge on many electrodes [11, 12]. An equation can be used to summarize it

$$\Delta Q_j = -q \left(\psi \left(x_f \right) - \psi \left(x_i \right) \right) \qquad (3)$$

where ΔQ_j is the change of charge induced by charge q on electrode j and x_i and x_f are separately the initial and final position of charge q. ψ is called weighting potential, calculated by the Laplace equation $\nabla^2 \psi = 0$. Each electrode has its own weighting potential distribution in the space. There are several prerequisites for this equation: the weighting potential of electrode j is set at 1, while on the other electrodes it is 0; the space charges are ignored [13]. According to Shockley-Ramo theory, assuming that the charge carriers all reach their corresponding electrodes, the change of the induced charge on a certain electrode from the movement of an electron-hole pair is

$$\Delta Q = -e_0 \left(0 - \psi \left(x_i \right) \right) + e_0 \left(1 - \psi \left(x_i \right) \right) = e_0 \qquad (4)$$

where e_0 is the electronic charge. It is obvious that the induced charge should only be relative with its parent charge. However, as the holes in CZT can merely move a short distance, the induced charge has to be corrected as

Table 2 Work function of different materials

Material	Al	Au	In	Pt	Cr	C	CZT
Work function (eV)	4.28	5.1	4.12	5.65	4.5	5.0	≈ 4.6

$$\Delta Q = -e_0 \left(\psi \left(x_i + \Delta x \right) - \psi \left(x_i \right) \right) + e_0 \left(1 - \psi \left(x_i \right) \right) \tag{5}$$

Thus, The induced charge is also position dependent. If the latter term is far larger than the former one, the contribution of hole to induced charge can be neglected, which is the principle of single polarity charge sensing technique. Many electrode structures basing on this technique has been invented, such as Frisch grid electrode, semi-hemisphere electrode, and pixelated electrode.

In this work, the conventional planar electrode is chosen because of its simple design and the short range of 5.486 MeV alpha particle in CZT (about 20 μm). As the cathode side faces the radiation source in experiment, the birthplace of electron-hole pair is close to cathode. Hence, the drift distance of hole is so short that holes can hardly be trapped.

Another factor taken into consideration when designing the detector is the contact formed by the electrodes and CZT. There are two types of metal-semiconductor contacts, Ohmic contact and Schottky contact. The characters of these two contacts are different, which could affect the performance of CZT detector. For Schottky contact, if the potential of the semiconductor is higher than metal, the leaking current will be quite low. Otherwise, the current will be strong. According to the principle of metal-semiconductor contact, if the work function of metal W_m is apparently larger than the one of semiconductor W_s, their contact will be Schottky. If $W_m \approx W_s$, the contact will be Ohmic [14]. Hence, proper electrode material should be chosen to form a supposed contact. The work function of several materials are listed in Table 2 [15, 16]. The CZT crystal might be asymmetric, so it is necessary to examine which electrode is more suitable to be selected as cathode. The Ohmic contact provides a convenient way of exchanging the electrodes with no need for remaking the electrodes. If the two contacts are all Ohmic, the value of the leaking current won't change after reversing the applied voltage, while the anode becomes a cathode and the cathode is transformed into an anode. It is reported that Au-CZT Ohmic contact has the lowest leaking current [17]. Thus, the material of the electrodes for the CZT detector is determined to be gold.

Finally, the detector is designed as a $22 \times 22 \times 0.7$ mm^3 CZT crystal covered with a 100-nm-thick Cr coating on the whole cathode surface, while two Au electrodes are separately on the two 22×22 mm^2 surfaces of the detector, seen in Fig. 2.

Fig. 2 Picture of the planar CZT detector. The crystal and cathode are beneath the square hollow made of Cr

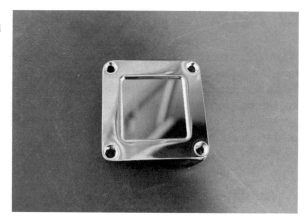

2.2 Electronics

The electronics is employed to transform the change of induced charge on anode into digital signal. It can be separated into several parts, including a high voltage module, a preamp, two amplifiers, two shaping filters, a trigger circuit, a sample/hold circuit, and an ADC board. The high voltage module is connected with the cathode and anode to provide bias voltage. The preamp is a charge-sensitive preamplifier, where the change of induced charge collected by anode is turned into voltage signal. Then the voltage signal is changed into two Gaussian waves with different amplitude and width by the couple of amplifiers and shaping filters. One of the Gaussian wave is 4 μs width, called "slow signal." Its height satisfies the requirement of ADC (analog-to-digital converter) board. Later it is turned into a 10-μs-wide square wave by the sample/hold circuit. The square wave has the ability of keeping a stable amplitude and is convenient for digital circuit to identify. The other Gaussian wave is 2 μs width, called "fast signal." The destination of the fast signal is the trigger circuit, consisting of a comparator, a single retriggerable monostable multivibrator, and a SPST (single-pole, single throw) trigger. If the amplitude of the fast signal exceeds the threshold of the comparator, a 2 μs width square wave will be created. Then the multivibrator will expand it to 10 μs width. Finally, this wide signal will open the SPST trigger for 10 μs, assuring the slow signal is absolutely received by subsequent ADC board. It should be mentioned that the work condition of the sample/hold circuit is controlled by the SPST trigger. When the SPST trigger is closed, the output of the sample/hold circuit is connected with ground, so slow signal can't be transferred to subsequent circuit. It is designed to prevent the ADC board from undesirable signal, for instance, electronic noise. The ADC board changes square wave into a number string. A MCU (micro control unit) and a Wi-Fi module are attached to the board. The digital signal could be delivered to a computer or a mobile phone by Wi-Fi or a USB data cable.

2.3 DAQ Program and Upper Computer Program

A DAQ (data acquisition) program and an upper computer program are developed to receive and process the data from the ADC board. The acquiring time is allowed to be set at a certain value. When acquiring a spectrum, the real-time counting rate, acquisition time, and a spectrum pattern are all shown in one window. The file format of saved spectrum is ".chn", which can be read by GammaVision, a widely used spectrum analyzer with authority.

3 Am-Pu α Particle Source Spectrum Acquisition

An Am-Pu alpha particle radiation source is utilized to test the performance of our CZT detector. The source contains two nuclides, Am-241 and Pu-239. In this part, the preparation for detection is introduced firstly, and then the detection procedure and results are shown and analyzed.

3.1 Linear Response Test

Before alpha particle detection, the linear response of the CZT detector has to be tested. Ideally, the spectrum channel is always proportional to the energy deposition in detector. However, many factors could render the relation between them nonlinear, such as polarization and signal saturation.

If the number of charge carriers trapped in the detector is sizable, a comparable electric field to the one from bias voltage will be created. This phenomenon is named polarization. Since the directions of the two fields are opposite, the charge carriers are slowed down, leading to longer drift time and higher rate of being trapped. As a consequence, the extra loss of charge carriers destroys the linear response. In our experiment, the activity of the radiation source we use is at a low level. The number of incident alpha particles per second is only a few tens. Therefore, the effect of polarization is negligible.

In our detecting system, signal saturation is possibly caused by the limitation of electronic components. If the amplitude of a signal transcends the threshold of an amplifier, the signal will be distorted and saturate. The part of the signal higher than the threshold will disappear; a platform flush with the threshold will take its place. Consequently, in the spectrum, the counts tend to accumulate at the channel representing amplifier's threshold rather than the amplitudes of original signals. In our electronics, considering the amplitude of signal from Am-Pu radiation source, to avoid signal saturation, the amplification factor was set below 12.

We only tested the linear response of the detecting system without the detector because alpha particle sources with different energy were unavailable. Instead,

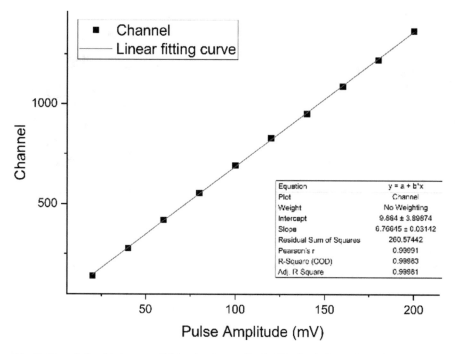

Fig. 3 The relationship between SPC and pulse amplitude. The line is linear fit result

a reference pulser (Canberra model 1407) was used to produce analog signal. The signal was transformed into charge by a 1 pF capacitor before entering the preamplifier. The anode was biased to 83 V where the signal-noise ratio (SNR) reaches a maximum. The relationship between signal peak channel (SPC) and pulse amplitude is shown in Fig. 3, confirming that the alpha detector has good linear response.

The amplitudes of the analog signals corresponding to 5.486 MeV alpha particle from Am-241 and 5.157 MeV from Pu-239 are calculated by the following equation:

$$U = \frac{E_\alpha}{E_I} \cdot C \tag{6}$$

in which U is the signal amplitude, E_α is the energy of alpha particle, $E_I = 4.64$ eV is the average ionization energy of CZT, and C is the capacity of the voltage-charge converter. The computation results are 189 mV for 5.486 MeV and 177 mV for 5.157 MeV, in the linear response interval of the detector and indicating that our detector is capable of the detection.

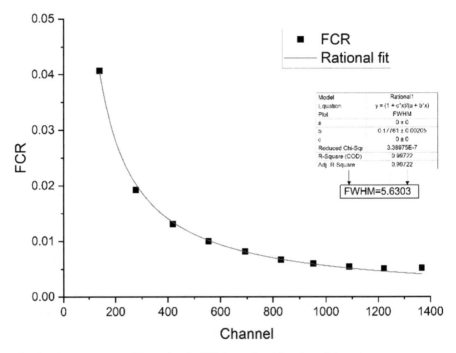

Fig. 4 FCR versus channel. Note that the FCR is a rational function of channel

3.2 Electronic Noise Measurement

The electronic noise was measured to evaluate the influence of electronics to the entire system. The measurement result is necessary for spectrum reconstruction. The anode was biased to 83 V, while the cathode was connected with ground. The test signal was also from the pulser Canberra model 1407. The FWHM (full width at half maxima) of the peak in the spectrum collected by the detector system was employed to estimate the system's electronic noise. The relationship between FWHM-channel ratio (FCR) and channel is shown in Fig. 4. FCR can directly present the effect of electronic noise on the performance of the detecting system, since its computational method is similar to that of energy resolution.

It was found that FCR is inversely proportional to channel. As a result, the noise expressed by FWHM is a constant in channel and equal to 5.6303. What's more, the influence of electronic noise decreases when the signal amplitude increases. The corresponding channel of 5.486 MeV and 5.157 MeV is between 1000 and 1200, where the FCR is around 0.05.

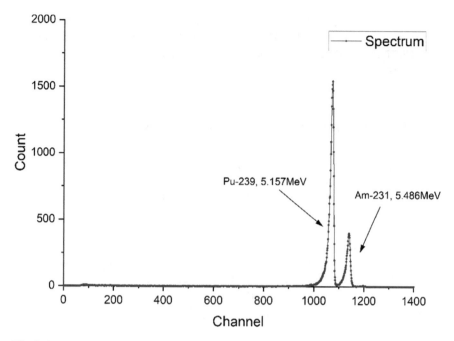

Fig. 5 Spectrum of Am-Pu α particle source from experiment, obtained when the anode is biased to 83 V and the voltage of the cathode is 0 V

3.3 Am-Pu α Particle Spectrum Acquisition

The Am-Pu alpha particle source is a wafer with a radius of 11.5 mm, and the radionuclides are on one side. Before detection, the surface of the radiation source was cleaned. Then the source was set parallel to the cathode of CZT detector with a distance of 1.4 cm. Since the source and the cover of detector are all electrically conductive, the source wasn't directly put on the detector. There was, therefore, a layer of polyethylene foam with a square hole placed between them. The source, the detector, and the electronics were put into an aluminum box, which served as electromagnetic screen. The electronics was powered by a portable charger outside the box. In order to prevent alpha particle's energy deposition in air, the box was laid into a vacuum oven, and the air pressure was lowered to 1 kPa (Fig. 5).

According to the spectra collected from experiment, the energy resolution achieved by the detector system was 1.47%@5.486 MeV and 1.32%@5.157 MeV. A better one, 1.22%@5.486 MeV and 1.17%@5.157 MeV, was achieved after biasing the cathode to −92 V and grounding the anode. It must be emphasized that biasing cathode will increase the risk of electric shock due to the electroconductive Cr coating, and that's why this design was abandoned. The energy resolution of Am-

241 is worse than that of Pu-239, whereas electronic noise should have less impact on the energy resolution at 5.486 MeV. The reason for this abnormal result might be the radiation source itself. As a prototype, the detector has shown a quite satisfying energy resolution.

3.4 Electron Mobility-Lifetime Product Measurement

Mobility-lifetime product is a significant parameter which characterizes the quality of semiconductor radiation detector [18]. In Eq. (2), it is shown clearly that there is a tight connection between CCE (charge collection efficiency) and mobility-lifetime product. CCE is equal to $\frac{Q_i}{N_e}$.

In this work, the ^{241}Am-^{239}Pu radiation source was aimed at the cathode of CZT detector to measure $\mu_e\tau_e$, while the bias voltage of anode was controlled by a high voltage power supply (Canberra model 3105). As the range of α particle in CZT is about 20 μm whereas the detector is far more thick, the contribution of holes to CCE can be ignored. The electric field E is the quotient of the bias voltage U divided by the thickness d. Therefore, Eq. (1) is simplified as

$$\text{CCE} = \frac{\mu_e\tau_e U}{d^2}\left(1 - \exp\left(-\frac{d^2}{\mu_e\tau_e U}\right)\right) \tag{7}$$

The peak channel of 5.157 MeV and 5.486 MeV is proportional to CCE. The fitting results of bias voltage and peak channel are shown in Fig. 6. The $\mu_e\tau_e$ is 2.74×10^{-7} m$^2 \cdot$ V^{-1} from the fitting result of Am-241, and 2.62×10^{-7} m$^2 \cdot$ V^{-1} from Pu-239.

4 Simulations of CZT Detector's Spectral Response

A series of simulations about the spectral response of the planar CZT detector were accomplished, consisting of interaction between α particles and CZT, charge carrier transport, and the electronic noise of CZT detector. As a result, a simulated spectrum of the Am-Pu source was built, which well matched the one from experiment. It is a symbol that we primarily have the ability of analyzing CZT detector's performance without empirical data.

4.1 Simulation of α Source and α-CZT Interaction

The alpha particle source and the interaction between alpha particle and CZT were simulated by a popular Monte Carlo simulation toolkit, Geant4 [19]. The configuration of the detector and the source is identical with the one in experiment. The initial position and momentum direction of the radioactive nuclide are determined

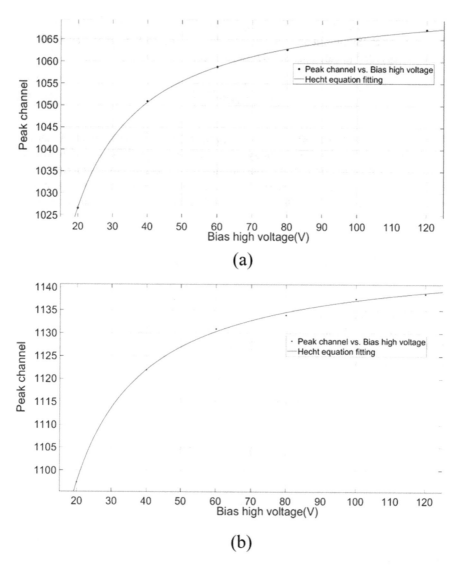

Fig. 6 Hecht fitting results. (**a**) The peak channel of 5.157 MeV vs. bias high voltage and the Hecht equation fitting curve. (**b**) The peak channel of 5.486 MeV vs. bias high voltage and the Hecht equation fitting curve

randomly. The energy of alpha particle in each event is selected basing on the branch ratio and the count rate of Am-241 and Pu-239 measured in experiment. We found that the number of steps in an incident event is always no more than three and most of the energy of alpha particle deposited in the first step. Therefore, to simplify data processing, only the deposited energy and a weighted energy depositing position were recorded for an incident event. These data would be used for spectrum construction and correction. The energy deposition and the intermediate position between the PreStepPoint and the PostStepPoint of the ith step, expressed as E_i and $\mathbf{r}_i$, were recorded to calculate the weighted energy depositing position

$$\mathbf{r} = \frac{\sum_{i=1}^{n} \mathbf{r_i} \cdot E_i}{\sum_{i=1}^{n} E_i} \tag{8}$$

and the total energy deposition $E = \sum_{i=1}^{n} E_i$, where n is the number of steps in an event. As all the deposited energy of α particle is transformed to e–h pairs in CZT and most of the deposition concentrates on a short length comparing to the entire incident range, $\mathbf{r}$ is appropriate to represent the initial position of the e–h pairs created by single α particle. There are 500,000 events generated per simulation.

4.2   Simulation of Charge Carrier Transport

The trapping of holes is severe in CZT such that it deeply influences the performance of CZT. To estimate the efficiency of the transport of charge carriers, a relative simulation has been achieved with the finite element methods (FEM) software, COMSOL MultiPhysics 5.5. The physics was given by two differential charge continuity equations [20].

$$\frac{dn}{dt} + \nabla \cdot (\mu_e n \nabla \varphi) - \nabla \cdot (D_e \nabla n) + \frac{n}{\tau_e} = f_n \tag{9}$$

$$\frac{dp}{dt} - \nabla \cdot (\mu_h p \nabla \varphi) - \nabla \cdot (D_h \nabla p) + \frac{p}{\tau_h} = f_p \tag{10}$$

$$D_{e,h} = \frac{k_B T}{q} \cdot \mu_{e,h} \tag{11}$$

where n and p are individually the density of electron and hole, $D_{e,h}$ corresponds to the diffusion coefficient, φ is the electric potential, $\tau_{e,h}$ is the lifetime of charge carriers in CZT, and $f_{n,p} = \delta(\mathbf{r} - \mathbf{r_0})\delta(t - t_0)$ represents a unit charge generated at position $\mathbf{r_0}$ and time t_0. Here the meaning of these differential equations will be explained. On the left side of Eq. (9), the second term $\nabla \cdot (\mu_e n \nabla \varphi)$ represents the drift of electron, the third term $- \nabla \cdot (D_e \nabla n)$ represents diffusion, and $\frac{n}{\tau_e}$ is the trapping term. It is similar in Eq. (10), the only difference is that the drift term of hole is inverse. The change of induced charge on anode, Q, caused by an e–h pair with charge q which generated at $t = 0$ and $\mathbf{r} = \mathbf{r_0}$ can be written as below [21].

$$Q_e = q \int_0^t dt' \int_\Omega d\Omega \cdot n\mu_e \nabla\phi\nabla\psi$$

$$Q_h = q \int_0^t dt' \int_\Omega d\Omega \cdot p\mu_h \nabla\phi\nabla\psi \tag{12}$$

$$Q = Q_e + Q_h$$

where Ω represents the whole volume of CZT crystal and ψ is the weighting potential of anode. The charge collection efficiency can be expressed as

$$\eta(r_0, t) = \frac{Q}{q} \tag{13}$$

To calculate CCE, the distribution of electric potential and anode weighting potential should be found out first. In this case, since the perturbation of electric potential resulting from the charge carriers induced by alpha particle is too small, it is assumed that the electric potential always obeys the following Laplace equation:

$$\nabla^2 \phi = 0 \tag{14}$$

Meanwhile, the potential at anode and cathode is set as 83 V and 0 V, separately. All the in- and outgoing fluxes through the other four detector surfaces are set as 0.

The weighting potential can also be computed by a Laplace equation

$$\nabla^2 \psi = 0 \tag{15}$$

with several boundary conditions. The weighting potential at anode is 1 and 0 at cathode, and there are no fluxes on the other surfaces.

In our simulation, the electric potential and weighting potential were solved firstly. Then the density of electron and hole were calculated and the boundary conditions were the same as described by Prettyman [21]. Finally, the CCE for each interaction position from Geant4 simulation results was obtained by the integrations in term of Eq. (12). However, it was soon found that the procedure was very time-consuming. Moreover, it's difficult to define a Dirac source in COMSOL. The source is always changed into an approximate distribution, which is not isotropic. Therefore, a Gaussian distribution was chosen to be the substitute. As the result, in this way, the solution of CCE might not be convincing.

For the reasons raised above, an efficient and accurate simulation method is required, and it is adjoint function in this work. First of all, write an operator L, and it functions as

$$L(n) = \frac{dn}{dt} + \nabla(\mu_e \cdot n \cdot \nabla\phi) - \nabla \cdot (D_e \nabla n) + \frac{n}{\tau_e} \tag{16}$$

The adjoint function of n is represented by n^*, and the adjoint operator is L^*. They are defined by [22].

$$\int_0^t dt' \int_\Omega d\Omega \left(nL^*\left(n^*\right) - n^*L(n)\right) = 0 \tag{17}$$

To give a suitable form of $L^*(n^*)$ satisfying the equation, it's better to analyze the terms of $L(n)$ one by one. For the term $\frac{dn}{dt}$, there is

$$\int_0^t dt' \int_\Omega n^* \frac{dn}{dt} d\Omega = \int_\Omega n^*nd\Omega|_0^t - \int_0^t dt' \int_\Omega n \frac{dn^*}{dt} d\Omega \tag{18}$$

Therefore, the adjoint form can be selected as $-\frac{dn^*}{dt}$.

For $\nabla(\mu_e \cdot n \cdot \nabla \phi)$, since $\nabla(\mu_e \cdot n \cdot \nabla \phi) = \mu_e \cdot \nabla n \cdot \nabla \phi + \mu_e \cdot n \cdot \nabla^2 \phi$ and $\nabla^2 \phi = 0$, there is

$$\int_0^t dt' \int_\Omega n^* \mu_e \nabla n \nabla \phi \, d\Omega = \int_{\partial \Omega} \mu_e \nabla \phi \, n^* n \cdot \hat{n} dS - \int_0^t dt' \int_\Omega n \mu_e \nabla n^* \nabla \phi \, d\Omega \tag{19}$$

the adjoint form is $-\mu_e \nabla n^* \nabla \phi = -\mu_e \nabla(n^* \nabla \phi)$.

For $\nabla \cdot (D_e \nabla n)$, as D_e is a constant,

$$\int_0^t dt' \int_\Omega n^* \nabla \cdot (D_e \nabla n) d\Omega$$

$$= \int_{\partial \Omega} D_e n^* \nabla n \hat{n} dS - \int_0^t dt' \int_\Omega D_e \nabla n^* \nabla n d\Omega \tag{20}$$

$$= \int_{\partial \Omega} D_e \left(n^* \nabla n - n \nabla n^*\right) \hat{n} dS + \int_0^t dt' \int_\Omega D_e n \nabla^2 n^* d\Omega$$

the adjoint form is $D_e \nabla^2 n^*$.

For $\frac{n}{\tau_e}$,

$$\int_0^t dt' \int_\Omega n^* \frac{n}{\tau_e} d\Omega = \int_0^t dt' \int_\Omega n \frac{n^*}{\tau_e} d\Omega \tag{21}$$

the adjoint term is $\frac{n^*}{\tau_e}$.

Finally, the adjoint operator can be expressed as

$$L^*\left(n^*\right) = -\frac{dn^*}{dt} - \nabla\left(\mu_e \cdot n^* \cdot \nabla \phi\right) - \nabla \cdot \left(D_e \nabla n^*\right) + \frac{n^*}{\tau_e}. \tag{22}$$

Notice that $L(n) = f_n = \delta(r - r_0)\delta(t - t_0)$. Assume $L^*(n^*) = g$, n^* is expected to be corresponding to CCE. It is found that

$$n^* (r_0, t_0) = \int_0^t dt' \int_\Omega n^* L(n) d\Omega = \int_0^t dt' \int_\Omega n^* f_n d\Omega$$

$$= \int_0^t dt' \int_\Omega n^* \delta (r - r_0) \, \delta (t - t_0) \, d\Omega = \int_0^t dt' \int_\Omega n L^* \left(n^* \right) d\Omega \tag{23}$$

$$= \int_0^t dt' \int_\Omega ng d\Omega = \int_{t_0}^t dt' \int_\Omega ng d\Omega$$

Comparing Eq. (23) with Eq. (12), it is found that if $g = \mu_e \nabla \phi \nabla \psi$ then $n^*(r_0, t_0) = \eta_e(r_0, t)$. It should be noted that obtaining n^* is a final value problem, while η_e is an initial value problem. To make them consistent, a new function was introduced as n^+ and used to replace n^* in Eq. (22).

$$n^+ (r_0, t) = n^* (r_0, t_0)$$

$$\frac{dn^+}{dt} = -\frac{dn^*}{dt_0} \tag{24}$$

It is apparent that n^+ is the CCE of electron. Finally, the adjoint equation of electron continuity can be written as

$$\frac{dn^+}{dt} - \nabla \cdot \left(\mu_e n^+ \nabla \varphi \right) - \nabla \cdot \left(D_e \nabla n^+ \right) + \frac{n^+}{\tau_e} = \mu_e \nabla \varphi \nabla \psi \tag{25}$$

while the boundary condition satisfies

$$\int_0^t dt' \int_{\partial \Omega} \mu_e \nabla \phi n^+ n \cdot \hat{n} dS + \int_0^t dt' \int_{\partial \Omega} D_e \left(n^+ \nabla n - n \nabla n^+ \right) \hat{n} dS = 0 \tag{26}$$

In the same way, the adjoint equation of hole is

$$\frac{dp^+}{dt} + \nabla \cdot \left(\mu_h p^+ \nabla \varphi \right) - \nabla \cdot \left(D_h \nabla p^+ \right) + \frac{p^+}{\tau_h} = \mu_h \nabla \varphi \nabla \psi \tag{27}$$

The initial value of n^+ and p^+ is 0 at $t = 0$. It should be noticed that in our work, the boundary condition was set differently from the one in citation [21]. Here, it is $\hat{n} \cdot \nabla D_h p^+ = -S_q p^+$ for anode and $\hat{n} \cdot \nabla D_e n^+ = -S_n n^+$ for cathode, where $S_n = 20 \, m \cdot s^{-1}$ and $S_p = 0.1 \, m \cdot s^{-1}$ are surface recombination velocity. Interestingly, we found in later research that the value of the surface recombination velocity doesn't influence the simulation result. The change of boundary condition was made because the original one, $p^+ = 0$ for anode and $n^+ = 0$ for cathode, would cause a sharp drop of electron CCE near anode and a steep cliff of hole CCE near cathode, while the truth is that the electron CCE near anode is the highest in the volume of CZT and the same for the hole CCE near cathode.

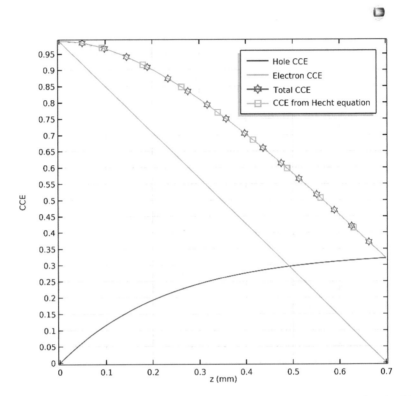

Fig. 7 CCE simulation results obtained after 4000 ns. The dark blue curve corresponds to the CCE of hole vs. z, the green curve to the electron CCE, red to the sum of electron and hole CCE, and the light blue curve is the CCE from the Hecht equation vs. z

Table 3 Properties for charge carriers applied in the simulation

Property	value
Temperature	300 K
Electron mobility μ_e	$2.315 \times 10^{-1} \ m^2 \cdot s^{-1} \cdot V^{-1}$
Hole mobility μ_h	$2 \times 10^{-3} \ m^2 \cdot s^{-1} \cdot V^{-1}$
Electron lifetime τ_e	$1.175 \times 10^{-6} \ s$
Hole lifetime τ_h	$1 \times 10^{-6} \ s$

Our simulation results of CCE are drawn in Fig. 7, corroborating that the contribution of hole CCE to the total CCE near the cathode is too little to take into consideration. The simulation is pretty successful, since the result is the same as predicted by the Hecht formula. The properties for charge carriers applied in the simulation are summarized in Table 3. The total CCE becomes stable after 4000 ns, which means the collection time won't be longer than 4 μs. Therefore, only the count rate more than $2.5 \times 10^5 \ s^{-1}$ can the signals overlap frequently.

4.3   Simulation of Electronic Noise

The electronic noise of the detector has been estimated in experiment, seen in Sect.3.2, and is a constant in channel. Since the unit of horizontal axis in simulated spectrum is keV, the electronic noise needs to be expressed as energy. The relationship between channel and energy is given by the peak channel of 5.486 MeV and 5.157 MeV measured in experiment (shown in Fig. 5), which can be expressed as

$$E = bC \tag{28}$$

where E is energy whose unit is keV, C is channel, $b = 4.796$ is a constant. The corresponding standard deviation of electronic noise N, σ, could be calculated by the following equation

$$\sigma = \frac{\mathrm{FWHM}}{2\sqrt{2\ln 2}} = \frac{\mathrm{FWHM}}{2.355} = \frac{N}{2.355} \tag{29}$$

Finally, the energy E' considering electronic noise can be expressed as

$$E' = E + b\sigma \cdot \mathrm{Gaussrand} = E + \frac{bN}{2.355} \cdot \mathrm{Gaussrand} \tag{30}$$

where E is the original energy deposition, N is 5.6303 in channel, Gaussrand is from a random number generator obeying standard normal distribution.

4.4   Spectrum Reconstruction

The simulation result of Geant4 provided the initial position of e-h pairs and original deposition energy of alpha particle in CZT detector. A spectrum can be drawn depending on these statistics to show the intrinsic energy resolution of the CZT detector. The position information was then be used in COMSOL to simulate corresponding CCE. Next, the CCE was multiplied with original deposited energy to obtain corrected energy, which illustrates the performance of CZT detector without the influence of electronics. Finally, the simulated electronic noise was added to the corrected energy. The result was utilized to reconstruct a spectrum reflecting the detecting system's energy resolution. The simulated spectra from each step mentioned above are shown in Fig. 8.

The simulated spectrum and the experimental spectrum match well, shown in Fig. 9. It should be informed that the experimental spectrum is the one obtained when the bias voltages of anode and cathode were set as 83 V and 0 V. FWHMs of the two spectra are evaluated by Origin, a software for data analysis. There is a clear shoulder on the simulated spectrum compared to the experimental one, which

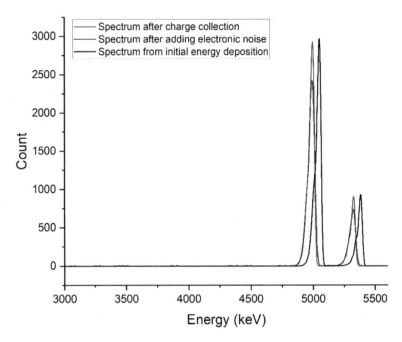

Fig. 8 The simulated spectra. Black: Spectrum drawn by the energy deposition of α particles from GEANT4. Red: Spectrum corrected by CCE. Blue: Spectrum after adding electronic noise

is caused by the crystal defects on the surface of CZT near cathode. In simulation, the influence of the defects is averaged in the bulk of crystal by the involvement of lifetime of charge carriers. Thus, the decrease of CCE resulting from the defects is unable to be absolutely shown in simulated spectrum, which shapes the shoulder.

4.5 Estimation of the Influence of 100 nm Cr Coating on the Performance of CZT Detector

As introduced in Sect. 2.1, there is a 100-nm-thick Cr coating on the whole cathode surface. Thanks to the hardness of Cr, the coating could provide a cover for the fragile Au electrode. Since the range of alpha particle in metal is only a few to tens of microns, the impact of the Cr coating on the energy resolution of the CZT detector needs to be evaluated. A detector model without the Cr coating was built, and a simulated spectrum of Am-Pu source was generated. By comparing it with the simulated spectrum shown in Fig. 9, we found that the peak of Pu-239 moves from 4984 keV toward 5024 keV after removing the coating, while the peak of Am-241 changes from 5315 keV to 5354 keV. What's more, there isn't apparent decrease in the number of incident events after adding the coating. The coating is too thin to stop the alpha particles. The energy resolution of Pu-239 and Am-241 is

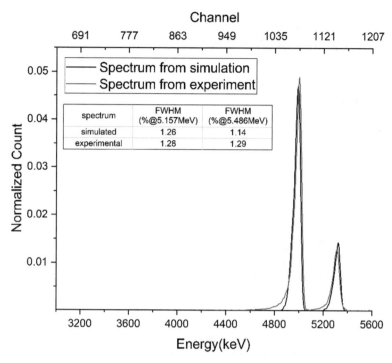

Fig. 9 The simulated spectrum with the experimental one. Black: The spectrum of Am-Pu source from simulation. Red: The spectrum of Am-Pu source from experiment. The horizontal axis of the simulated spectrum is at bottom and represents energy, while the one of experimental spectrum is at top representing channel

improved to 1.02%@5157 keV and 0.93%@5486 keV. The statistics mentioned in this subsection are read out by Origin. The two spectra are shown in Fig. 10.

Although in this work the coating has an acceptable effect on the detector's performance, it doesn't means that it can be overlooked under any conditions. If the energy of the incident alpha particles ranges from a few tens of keV to a few hundreds of keV, which is usual in neutron detection, the coating could severely decrease the detector's efficiency and cause lower signal-noise ratio. So could the Au electrodes. Therefore, when manufacturing the detector, the coating and electrodes should be machined as thin as possible.

5 Conclusion

An α particle detecting system basing on a planar CZT detector has been designed and manufactured in this work. The energy resolution achieved by the system is 1.47%@5.486 MeV and 1.32%@5.157 MeV. A series of corresponding simulations

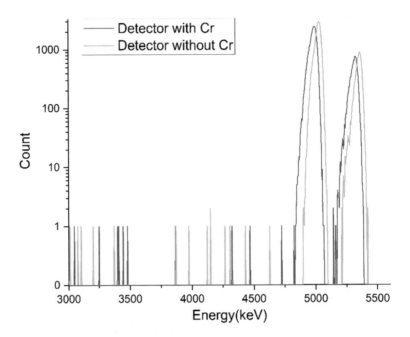

Fig. 10 Red: Simulated spectrum from the detector with Cr coating. Green: Simulated spectrum from the detector without Cr coating. The vertical axis is in log scale

has also been done, including the interaction between alpha particle and CZT, charge carrier transport and electronic noise. A simulated spectrum has been reconstructed relying on the simulation results and matched the experimental one well. The simulation model we built is supposed to provide a better design for the detecting system in the future.

References

1. Yang, M. T. (2011). The present status and develop current of α particle detect instrument. *Nuclear Electronics & Detection Technology, 31*(11), 1198–1201. (in Chinese).
2. Morishita, Y., Yamamoto, S., Izaki, K., et al. (2014). Performance comparison of scintillators for alpha particle detectors[J]. *Nuclear Instruments and Methods in Physics Research Section A: Accelerators, Spectrometers, Detectors and Associated Equipment, 764,* 383–386.
3. Venos, D., Srnka, D., Slesinger, J., et al. (1995). Performance of HPGe detectors in the temperature region 277 K[J]. *Nuclear Instruments and Methods in Physics Research Section A: Accelerators, Spectrometers, Detectors and Associated Equipment, 365*(2–3), 419–423.
4. Kotina, I. M., Derbin, A. V., Morozov, V. F., et al. (1992). Detection of charged particles using heterostructures of crystalline p-silicon and hydrogenated amorphous carbon[J]. *Diamond and Related Materials, 1*(5–6), 623–625.
5. He, Y., Liu, Z., McCall, K. M., et al. (2019). Perovskite CsPbBr3 single crystal detector for alpha particle spectroscopy[J]. *Nuclear Instruments and Methods in Physics Research Section A: Accelerators, Spectrometers, Detectors and Associated Equipment, 922,* 217–221.

6. Xu, Q., Mulligan, P., Wang, J., et al. (2017). Bulk GaN alpha-particle detector with large deple-tion region and improved energy resolution[J]. *Nuclear Instruments and Methods in Physics Research Section A: Accelerators, Spectrometers, Detectors and Associated Equipment, 849,* 11–15.

7. Amman, M. S., Lee, J. S., & Luke, P. N. (2001). Alpha particle response characterization of CdZnTe[C]. Hard X-Ray and Gamma-Ray Detector Physics III. *International Society for Optics and Photonics, 4507,* 1–11.

8. Wang X. Principle and experimental characteristics of pixellated CdZnTe detector for nuclear radiation, Ph.D. Thesis, Chongqing University; 2013. (in Chinese).

9. Hecht, K. (1932). Zum Mechanismus des lichtelektrischen Primrstromes in isolierenden Kristallen[J]. *Zeitschrift fr Physik, 77*(3–4), 235–245. (in German).

10. 王莹, 王凯, 曲延涛, 马吉增. 平面型CZT探测器对中低能γ射线的响应[J]. 中国原子能科学研究院年报. 2007(00):356. (in Chinese).

11. Shockley, W. (1938). Currents to conductors induced by a moving point charge. *Journal of Applied Physics, 9,* 635.

12. Ramo S. Currents induced by electron motion. Proceedings of the I.R.E. 1939, p. 584.

13. He, Z. (2001). Review of the Shockley–Ramo theorem and its application in semiconductor gamma-ray detectors[J]. *Nuclear Instruments and Methods in Physics Research Section A: Accelerators, Spectrometers, Detectors and Associated Equipment, 463*(1–2), 250–267.

14. Liu, E. K., Zhu, B. S., & Luo, J. S. (2017). *The physics of semiconductors* (7th ed.). Publishing House of Electronics Industry.

15. Michaelson, H. B. (1977). The work function of the elements and its periodicity[J]. *Journal of Applied Physics, 48*(11), 4729–4733.

16. Liang, X. Y., Min, J. H., Chen, J., et al. (2012). Metal/semiconductor contacts for schottky and photoconductive CdZnTe detector[J]. *Physics Procedia, 32,* 545–550.

17. Nemirovsky, Y., Ruzin, A., Asa, G., et al. (1997). Study of contacts to CdZnTe radiation detectors[J]. *Journal of Electronic Materials, 26*(6), 756–764.

18. Uxa, S., Grill, R., & Belas, E. (2013). Evaluation of the mobility-lifetime product in CdTe and CdZnTe detectors by the transient-current technique[J]. *Journal of Applied Physics, 114*(9), 094511. https://doi.org/10.1063/1.4819891

19. Agostinelli, S., Allison, J., Amako, K., et al. (2003). GEANT4—a simulation toolkit[J]. *Nuclear Instruments and Methods in Physics Research Section A: Accelerators, Spectrometers, Detectors and Associated Equipment, 506*(3), 250–303.

20. Kolstein, M., Ario, G., Chmeissani, M., et al. (2014). Simulation of charge transport in pixelated CdTe[J]. *Journal of Instrumentation, 9*(12), C12027.

21. Prettyman, T. H. (1999). Method for mapping charge pulses in semiconductor radiation detectors[J]. *Nuclear Instruments and Methods in Physics Research Section A: Accelerators, Spectrometers, Detectors and Associated Equipment, 422*(1–3), 232–237.

22. Bell, G. I., & Glasstone, S. (1970). *Nuclear reactor theory[R].* US Atomic Energy Commis-sion.

CdZnTeSe: Recent Advances for Radiation Detector Applications

Utpal N. Roy and Ralph B. James

1 Introduction

The availability of radiation detection technology operating at room temperature with energy dispersive spectroscopic capability in the X- and gamma-ray energy range has opened a wide range of applications in homeland security, nonproliferation, medical imaging, astrophysics, and high-energy physics [1–9]. To address the growing demand of end users for these applications, enormous R&D effort has been devoted worldwide for more than three decades to develop effective detection materials at a lower production cost. The requirements of different candidate materials used for radiation detector applications operating at room temperature are very stringent and multi-pronged. For enhanced stopping power of high-energy gammas, the materials should contain at least one primary element with a high atomic number. In addition, the constituent elements of the detector materials should not contain radioactive isotopes to eliminate unwanted intrinsic radioactive events. The bulk resistivity of the material should be $>10^{10}$ ohm-cm at room temperature to lower the leakage current in the dark state and reduce thermal noise. To achieve this required high-resistivity, the bandgap of the chosen material needs to be greater than ~1.5 eV. Unlike other conventional semiconductor detectors for measuring charged particles, gamma-ray radiation detectors demand relatively large thicknesses for better detection efficiency, especially for energies greater than about 50 keV. Thus, to ensure the complete collection of photo-generated charge, the material should have an adequately high mobility-lifetime product ($\mu\tau$). For materials with higher $\mu\tau$ values for electrons, the allowable thickness for functional high spectroscopic-performance devices can be larger. In order to achieve a uniform internal electric

U. N. Roy (✉) · R. B. James
Savannah River National Laboratory, Aiken, SC, USA
e-mail: Utpal.Roy@srnl.doe.gov

© The Author(s), under exclusive license to Springer Nature Switzerland AG 2023
L. Abbene, K. (Kris) Iniewski (eds.), *High-Z Materials for X-ray Detection*,
https://doi.org/10.1007/978-3-031-20955-0_8

field in the detector, the material needs to possess very low concentrations of defects that trap electrons or holes. These stringent characteristics have been a huge challenge to the crystal growth community to prudently develop the material with the right properties. To attain a lower cost of production, it is highly desirable to grow the material from a congruent melt at a relatively low temperature and be capable of scaling up to higher production via growth of large-volume ingots with reasonably low intrinsic and extrinsic defect concentrations. Further reductions in the extrinsic defects should be relatively easy using existing purification techniques, while reducing the intrinsic defects for a particular material/compound is likely more difficult. The chemical and physical stability of the metal-to-semiconductor junction is also an important aspect for prolonging the lifetime of a detector and avoided the onset of polarization.

Over the last three decades, only a handful of semiconductor materials have been discovered with the desired attributes. CdTe, CdZnTe (CZT), HgI_2, and TlBr are the most prominent ones [10–14]. In the recent years, perovskites have evolved into a promising family of materials for room-temperature semiconductor detector (RTSD) applications for high-energy radiation [15]. Johns et al. [16] listed a comprehensive list of possible radiation detector materials in a review article. Each of these materials however does suffer from unique issues associated with the difficulty of crystal growth, high concentrations of defects, and device stability.

The ternary material CZT remains the gold standard and has dominated the commercial sector for three decades, despite intense search for the ideal material globally. It should be noted that CZT is a by-product of CdTe. CdTe has multiple uses and has long been known as a substrate material for IR and night vision applications, and since the 1970s, it has also been regarded as a potential material for nuclear radiation detector applications [17]. In later years, Zn was added to the CdTe matrix for better lattice matching to HgCdTe epitaxial films for night vision applications, while Zn was added to CdTe to increase the bandgap of the material to achieve higher resistivity and prevent polarization effects associated with Cl-doped CdTe radiation detectors. Most of the CZT used today for radiation detector applications contains ~10 atomic % of Zn in the CdTe matrix. CZT-based radiation detector applications gained serious momentum beginning in the early 1990s after the onset of industrial commercialization. Initially, the detector-grade commercial CZT was grown from the high-pressure Bridgman technique [18, 19]. Around the same time in 1990, Triboulet et al. [20] demonstrated the growth of CdTe and ZnTe by cold traveling heater method (CTHM), and in 1994 the same group successfully grew CZT by the CTHM technique. In the later years, CTHM became commonly known as the traveling heater method (THM). A Te-rich melt was used for the THM growth technique. The major advantage of using a Te-rich melt in the THM technique is the lower temperature for growth. Thus, using the THM technique, CZT ingots can be grown at temperatures well below its melting point. The THM technique offers several other advantages as compared to melt-grown methods. Because of the lower temperature growth, the ingots possess less defects and thermal stress. The THM-grown CZT ingots were reported to show better axial and radial compositional homogeneity with a higher purity as compared to melt-grown ingots [21]. Redlen Technologies was founded in 1999 for commercial production of detector-grade CZT employing the THM growth technique. Because of the demonstrated advantages and successes of the THM technique, other companies began using the THM growth method to produce detector-grade CZT materials. Over time the THM technique has proven to be the most viable technique to grow CdTe and CZT commercially, and very large diameter ingots (up to 10 cm) can be produced [22]. However, even after decades of intense effort to improve the growth process, CZT material still possesses several shortcomings that severely limit its widespread deployment, especially for large-volume detectors. Although these shortcomings have been very difficult to fully resolve, the quality of CZT material has improved considerably since its early years, and the production cost has steadily decreased. The presence of high concentrations of sub-grain boundary networks and Te inclusions remains long-standing issues associated with CZT radiation detectors. These material defects act as charge trapping and recombination centers and severely affect the spatial uniformity of the charge-transport characteristics of the material resulting in limitations on the maximum thickness of high-performing detectors. Incremental enhancement in detector performance has been demonstrated as the size and concentration of Te inclusions has decreased through modifications in the growth and post-processing conditions. In addition to these extended structural defects, CZT suffers from axial and radial compositional inhomogeneity due to the non-unity segregation of Zn in the CdTe matrix [23], which heavily hampers the overall yield of detector-grade material particularly for thick detectors. In addition, the compositional gradient is known to impose considerable strain in the CZT ingot [6, 24]. This article will discuss the sub-grain boundary networks and Te inclusions in CZT and their effects on device performance. The successful mitigation of these defects by adding selenium in the CZT matrix will be discussed. An overview of recent developments of the new quaternary material $Cd_{1-x}Zn_xTe_{1-y}Se_y$ (CZTS) and optimization of the composition and charge-transport characteristics will also be discussed as a potential new next-generation detector material operable at room temperature.

2 Te Inclusions and Sub-grain Boundaries and Their Networks in CZT

Most of the present-day as-grown CZT materials contain high concentrations of Te inclusions and sub-grain boundaries and their networks, irrespective of the growth technique. Te-rich secondary phases in the CdTe/CZT matrix are generally defined by the way they are created during either the growth or subsequent cooling process, (i) inclusions and (ii) precipitates [7]. Inclusions are typically considered to be generated due to a morphological instability at the growth interface [7, 25–27], while precipitates are generated during the cooling process due to the retrograde solubility of tellurium [7] in CZT. The size of the Te precipitates is in the range of 10–50 nm[7], and the size of the inclusions can vary between 1 and 20 μm.

Inclusions can even be larger than 50-μm diameter. Due to the scattering of infrared light by below bandgap illumination of CZT, IR transmission microscopy can generally detect inclusions with sizes greater than 1–2 μm. However, the precipitates generated due to the retrograde solubility are too small to detect by IR microscopy. They can migrate during the post-growth cooling process under the influence of a thermal gradient, which causes them to coagulate with other precipitates or inclusions, ultimately forming large inclusions with size greater than 1 μm. Thus, even if the formation of Te-rich inclusions due the instability of growth interface is avoided, larger Te-rich secondary phases might still be present in the CdTe/CZT matrix.

Te inclusions are responsible for creating charge trapping centers and severely hampering the charge-transport properties, which eventually degrade the detector performance. Additionally, Te inclusions are reported to introduce a trap level at 1.1 eV above the valence band [28]. Carini et al. [29] demonstrated direct evidence of the charge loss due to the presence of Te inclusions. The mobility-lifetime product for electrons $[(\mu\tau)_e]$ in CZT can vary over a larger range of between (0.2–20) $\times$ 10^{-3} cm^2/V for the regions with Te inclusions compared to the clear regions that are relatively free from large inclusions [29]. The Te inclusions are in general randomly distributed within the bulk of CZT single-crystal material; however, the concentration is typically not uniform. The distribution of these localized defects in CZT matrix imposes severe spatial charge-transport inhomogeneity, which eventually broadens the photopeaks and degrades the spectroscopic performance. The adverse effect of Te inclusions on the energy resolution of CZT detectors was modeled for a 15-μm-thick Frisch grid detector [30]. The effect directly depends on the size, concentration, and thickness of the detectors [30]. For example, the number of Te inclusions with a size ~5 μm that can be tolerated is up to ~2 $\times$ 10^5 cm^{-3} to achieve an energy resolution of less than 1% at 662 keV. In contrast, a concentration of only ~10^3 cm^{-3} or less of Te inclusions with a size ~20 μm is allowable to keep the resolution less than 1% at 662 keV for 15-μm-thick Frisch grid detector, assuming a uniform internal electric field of strength ~2000 V/cm. The Te inclusions however can successfully be eliminated from the CZT matrix by post-growth annealing, and it is a common practice in the CZT detector manufacturing process. Unfortunately, the process is known to produce large (50–100 times larger than the size of the Te inclusion before annealing) star-like defects (punching defects) at the locations of relatively large Te inclusions. The star-like defects are invisible in an IR transmission measurement, but they can be detected by charge-collection mapping, preferential etching, and X-ray topographic experiments. Depending on their concentration and size, these star-like defects can also severely reduce the charge-transport properties [31] and the device performance [32]. Furthermore, Te inclusions are known to be surrounded by a dense field of dislocations, and those dislocations are expected to remain after removing the inclusions by thermal annealing [33].

Besides high concentration of Te inclusions, various structural defects are also present in CdTe/CZT, which adversely affect the charge-transport properties of the material and the performance of fabricated devices. The presence of sub-grain boundaries and their networks is perhaps the most serious performance-limiting defects causing serious degradation of detector quality and broadening of the pulse height. Most II–VI compounds are prone to the generation of strain and extended defects in as-grown ingots as a consequence of the high ionicity of the Cd-Te chemical bond of 0.55 [6]. This residual strain in CdTe ingots leads to the formation of sub-grain boundaries and their networks. The ionicity is inversely proportional to the energy of creation for dislocations and for stacking faults [6]. Thus, the higher the ionicity, the smaller the energy for formation of vacancies, dislocations, and stacking faults [6]. Higher ionicity and the poor thermophysical properties of CdTe have limited the improvement of the material quality over the years and made it difficult to grow large-diameter (>75–100 mm) ingots that are needed for production scaleup of large-volume high-quality detectors with less material defects.

The ternary material CZT came up as a consequence of the desire to improve the properties of CdTe. Adding Zn in CdTe matrix offered several advantages for different applications. For example, adding Zn in CdTe provides better lattice matching substrates to HgCdTe epilayers for IR or night vision applications. For radiation detector applications, the bandgap is enhanced to help satisfy the required resistivity of over 10^{10} ohm-cm at room temperature for manufacturing low-noise nuclear detectors with improved performance.

Perhaps the most beneficial reason for adding Zn in CdTe is the higher binding energy of the Zn-Te bond and lower ionicity. Thus, the addition of Zn in CdTe was reported to reduce the dislocation density and sub-grain boundaries and enhance the micro-hardness due to the lattice hardening effect caused by the introduction of Zn [6, 34–36]. Although some lattice hardening effect was observed after addition of Zn in the CdTe matrix, CZT still comprises a relatively high density of sub-grain boundaries and their networks. The low-angle sub-grain structures of size ~50–1000 μm tilted/misoriented by about 10–100 arc-sec [7] form the sub-grain boundary, and the interconnectivity of sub-grain boundaries is often called the sub-grain boundary network. The sub-grain boundaries are basically clusters of high-density dislocations decorated inside the sub-grain boundaries/walls. In most cases, these sub-grain boundaries and dislocation walls originate due to thermoelastic stress and dislocation polygonization immediately after solidification near the growth interface. The distribution of these sub-grain boundaries is highly nonuniform, and it is often observed that some portions of a single grain are free from sub-grain boundaries just adjacent to a highly decorated sub-grain boundary network [30]. In addition, most often the sub-grain boundaries are found to be highly decorated with Te inclusions and extrinsic impurities [7, 30], responsible for creating charge-trapping centers. The sub-grain boundaries and their network are invisible under IR transmission microscopy; however, they can be revealed by preferential etching of polished surfaces as well as X-ray topographic image taken in the reflection mode. Figure 1 (top) shows a preferentially etched CZT sample surface with dimensions of 9×3 mm^2 and the corresponding X-ray topographic image (bottom of Fig. 1). The dark lines in the cellular structure are arrangements of dislocations along the sub-grain boundary in the CZT sample. The density of sub-grain boundaries is particularly high for this sample. The dark spots in the

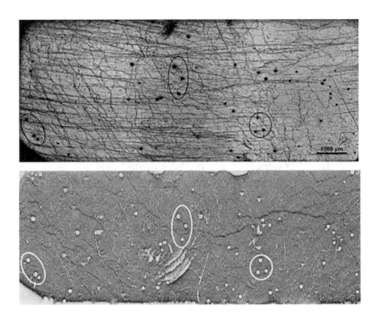

Fig. 1 Sub-grain boundaries and prismatic dislocations, usually generated around Te inclusions after post-growth annealing, are apparent on the surface of a 9×3 mm^2 area of the crystal treated with the Nakagawa-etching technique (top) and with X-ray diffraction topography (bottom). The circles denote the prismatic dislocations defects seen in both images. (Taken from Bolotnikov et al. [30])

ellipses denoted in Fig. 1 (top) are commonly known as punching defects, which form on sites of Te inclusions that have been removed by post-growth annealing. The X-ray topographic image of the same sample also shows the presence of sub-grain boundaries and their network with the appearance of dark and bright lines. The punching defects appear as white spots on the X-ray topographic images as highlighted inside the ellipses. The presence of sub-grain boundaries and their networks are possibly the most concerning extended defects in CZT material among those affecting the detector performance.

The sub-grain boundaries and their networks were found to be present in CZT samples grown by various techniques. Bolotnikov et al. [30] reported the presence of sub-grain boundaries and their networks in CZT samples by X-ray topographic measurements produced from seven different vendors across the globe. A drastic improvement of CdTe/CZT is limited due to its inherent poor thermophysical material properties.

Due to the random distribution of Te inclusions and sub-grain boundary networks, different pulse heights are produced at different locations of the detector sample, which blurs the resulting photopeak of the detector after integrating detector responses over the entire active volume. The performance degradation becomes more prominent with increased thickness. The effect of sub-grain boundaries on the performance of Frisch grid detectors was reported to be very prominent as compared for two different detector samples of the same dimensions [30]. Both the detectors showed the presence of a sub-grain boundary network, while the sample with the higher density of major sub-grain boundaries displayed larger broadening of the photopeaks. The presence of sub-grain boundaries in the CdTe-alloy family creates a serious issue for medical imaging applications as well. Buis et al. [37] demonstrated that the image of a cowry shell taken with a CdTe pixelated detector contained various lines corresponding to dislocation walls and sub-grain boundaries and needed an extra effort to eliminate after a flat-field correction.

Thus, the presence of high concentrations of Te inclusions and sub-grain boundary networks severely impairs the production of high-quality detectors at a lower cost. There is an urgent need to develop material free from such performance-limiting defects to achieve spatial charge-collection uniformity to further enhance the detector performance and overall yield of ingots.

3 Effects of Se Addition to the CdZnTe Matrix

The lesson learned from intense research over the past 30 years is that the unique phase diagram of CZT with its retrograde solubility of Te and its poor thermophysical properties make it difficult to produce as-grown ingots free from Te inclusions and sub-grain boundaries. To improve the crystallinity and increase the volume of defect-free material, another approach has been pursued via isoelectronic doping as an effective means for solid-solution hardening. The infrared substrate community observed a profound reduction of sub-grain boundaries and their networks after doping CdTe with as little as 0.4% (atomic) selenium, and they also showed that solid solution hardening with Se is more effective than with Zn (see, e.g., Refs. [7] and [38]). Tanaka et al. demonstrated the advantage of Se doping of the CZT matrix for substrate applications [39]. At an early stage of CZT detector development (in 1994), Fiederle et al. [40] reported a better mobility-lifetime ($\mu\tau$) product for electrons (4.2×10^{-4} cm^2/V) and a superior charge-collection efficiency for Bridgman-grown CdTe$_{0.9}$Se$_{0.1}$ (CTS) crystals compared to CZT. In recent years the addition of selenium in the CdTe matrix was also shown to greatly reduce the sub-grain boundaries and produce crystals free from sub-grain boundary networks while also possessing a reduced concentration of Te-rich secondary phases and a high compositional homogeneity [41–43]. The resulting ternary material CdTeSe, however, was not useful for detector applications due to a higher leakage current. Although the bandgap of CdSe is higher than CdTe, the bandgap of the quaternary CTS was reported to be less than CdTe due to a lattice disorder effect [44]. As a result, the required resistivity could not be attained for room-temperature operation.

To attain the required resistivity for realizing room-temperature radiation detector applications, the bandgap must be enhanced. Thus, the feasibility of adding selenium into the Cd$_{0.9}$Zn$_{0.1}$Te compound was explored. As in the case for CTS, the addition of a small amount of selenium in the CZT matrix improved the material quality significantly and successfully resolved some of the long-standing issues pertaining to CZT. The resulting quaternary compound Cd$_{1-x}$Zn$_x$Te$_{1-y}$Se$_y$ (CZTS)

Wafer periphery, touching the ampoule wall

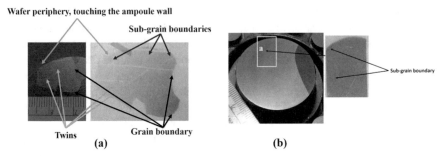

Sub-grain boundaries

Sub-grain boundary

Twins   Grain boundary
        (a)   (b)

Fig. 2 (a) Optical photograph of the grain (left) and the corresponding X-ray topographic image of the grain of a THM-grown $Cd_{0.9}Zn_{0.1}Te_{0.93}Se_{0.07}$ wafer (Taken from Roy et al. [45]) and (b) Photograph of a 4-cm diameter vertical Bridgman-grown $Cd_{0.9}Zn_{0.1}Te_{0.93}Se_{0.07}$ lapped wafer and the X-ray topographic image of the region denoted by the white rectangle. (The white rectangle and the corresponding X-ray topographic images are not to scale). (Taken from Roy et al. [46])

was found to be completely free from a sub-grain boundary network with occasional occurrence of isolated sub-grain boundaries for both THM- and Bridgman-grown samples [45, 46]. Figure 2a shows an optical photograph of a grain (left) from a 2-inch $Cd_{0.9}Zn_{0.1}Te_{0.93}Se_{0.07}$ wafer cut perpendicular to the growth axis for a THM-grown ingot and the corresponding X-ray topographic image of the same grain shown in the right side of the figure. In order to evaluate the effectiveness of selenium as a solution hardening agent, the ingot was naturally cooled to room temperature after the completion of ingot growth to force the introduction of thermal stress. It is to be noted that no lattice distortion was observed near the periphery of the wafer touching the ampoule wall, as indicated by the blue arrow, depicting the absence of any stress introduced from the ampoule wall. Figure 2b shows a vertical Bridgman-grown $Cd_{0.9}Zn_{0.1}Te_{0.93}Se_{0.07}$ wafer (left) cut perpendicular to the growth axis and an X-ray topographic image of the portion denoted by "a" on the wafer. Even though the Bridgman-grown ingots undergo higher thermal stress as compared to the THM-grown ingots, no lattice distortion was observed near the periphery of the wafer touching the ampoule wall. The material was observed to be free from sub-grain boundary network. Furthermore, the undistorted twins in the X-ray topographic image, as indicated by the green arrows in Fig. 2a, provide evidence that the material is free from any thermal stress. The X-ray topographic observations on the CZTS samples evidently confirm the effectiveness of selenium as a solid solution hardening agent and agrees well with previous findings [7, 38]. The addition of selenium was also found to be very effective in reducing the concentrations/size of Te inclusions as reported by various groups of researchers [45, 47–50]. The addition of selenium in the CZT matrix offers several other advantages over conventional CZT, including enhanced axial and radial compositional homogeneity with higher microhardness [46, 48, 51–55]. A uniform composition for $Cd_{0.96}Zn_{0.04}Te_{0.84}Se_{0.16}$ ingots grown by the vertical Bridgman technique was reported by Chang et al. [55]. Thus, selenium was found to have significant influence on modifying the Zn segregation in the CZT matrix. The substantial reduction in the density of deep trap levels has also been reported for selenium-containing CdTe base ternary and

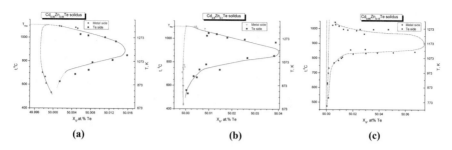

Fig. 3 T–X projection of the $Cd_{1-x}ZnxTe$ solidus for $x = 0.05, 0.1$ and 0.15. (Taken from Guskov et al. [67])

quaternary materials, viz., in CdTeSe and CdZnTeSe as compared to conventional CZT [28, 56, 57]. The new quaternary material was also reported to outperform CZT for high flux detectors for medical imaging applications [57]. Hence, selenium acts as an effective element in the CZTS matrix to mitigate several long-standing issues pertaining to CZT, and CZTS exhibits superior material quality. Due to the superior nature of the new quaternary compound, CZTS material has attracted considerable attention and is being actively studied in depth by various groups to better understand the material and optimize the growth process [28, 47, 48, 53, 54, 56–66].

Being a quaternary compound, the optimization of the composition of $Cd_{1-x}Zn_xTe_{1-y}Se_y$ is non-trivial. In addition, the complex nature of the phase diagram of $Cd_{1-x}Zn_xTe$ and the bandgap variation with increasing selenium in the CZT matrix make it a multi-faceted problem. Projection of the $Cd_{1-x}Zn_xTe$ solidus onto the T–X (temperature-composition) plane, as determined by Guskov et al. [67], is shown in Fig. 3a–c for different Zn compositions of 5 at. %, 10 at. %, and 15 at. %. As shown in Fig. 3, the general shape of the solidus is similar to the phase diagram of CdTe with a considerable higher retrograde Te solubility. The bulge on the Te side becomes more pronounced with increasing ZnTe content in the solid solution. Thus, due to the higher retrograde solubility of Te with increased Zn content in the $Cd_{1-x}Zn_xTe$ composition, higher concentrations of Te content (in the form of Te-rich secondary phases) are expected in the ingots grown with higher Zn content.

On the other hand, the addition of selenium in CdTe matrix imposes another disadvantage. Although the bandgap of CdSe is higher than CdTe, the bandgap of the resulting ternary CdTeSe decreases below the value of CdTe with increased Se concentration up to ~40% Se in the composition due to lattice disorder [44]. The bandgap variation of the quaternary compound $Cd_{1-x}Zn_xTe_{1-y}Se_y$ can be estimated following the empirical formula reported by Brill et al. [68]. From the empirical formula (Eq. 1), it is evident that the bandgap of the quaternary material increases with increasing concentrations of Zn while decreasing with increasing concentrations of Se.

$$E_g(x, y) = 1.511 - 0.54y + 0.6x \quad (x, \ y \leq 0.10) \quad (1)$$

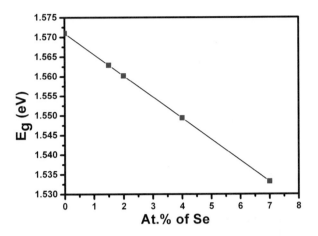

Fig. 4 Bandgap of
$Cd_{0.9}Zn_{0.1}Te_{1-y}Se_y$ with
selenium concentration.
(Taken from Roy et al. [69])

Hence, an increase of selenium and zinc content in the $Cd_{1-x}Zn_xTe_{1-y}Se_y$ quaternary crystal imposes an opposite effect on changes in the bandgap for different alloy compositions. Additionally, the increased Zn concentration boosts the Te retrograde solubility resulting in an increased concentration of Te precipitates for higher Zn content, while increasing the selenium concentration in the matrix results in a decrease in the bandgap energy. The reduced bandgap energy ultimately compromises the resulting resistivity of the material for radiation detector uses. To restrict the concentrations of the secondary phases (Te precipitates and inclusions) in the resulting quaternary material, the reported optimum concentration for CZT of 10 atomic-% zinc was kept constant, and selenium was added to the $Cd_{0.9}Zn_{0.1}Te$ matrix for optimizing the composition [69]. Various compositions with selenium concentrations ranging from 1.5 to 7 atomic % were investigated for the quaternary compound $Cd_{1-x}Zn_xTe_{1-y}Se_y$ while keeping $x = 0.1$ for all compositions. The THM technique was used to grow the ingots for this study. The resulting bandgap of the $Cd_{0.9}Zn_{0.1}Te_{1-y}Se_y$ was found to be lower than conventional $Cd_{0.9}Zn_{0.1}Te$ detector-grade material, consistent with expectations.

Figure 4 shows the bandgap variation with increasing concentrations of Se (up to 7 at. %) in the $Cd_{0.9}Zn_{0.1}Te$ matrix, calculated from Eq. 1. Due to the reduced bandgap energy with increased selenium concentration, the aim of the optimization process was to determine the minimum amount of selenium required to obtain a dramatic reduction in the sub-grain boundaries in $Cd_{0.9}Zn_{0.1}Te_{1-y}Se_y$ with a relatively low concentration of Te-rich secondary phases while keeping the resistivity high enough for realizing room temperature operation [69].

To study the presence of sub-grain boundaries and their network in the as-grown CZTS samples, white beam X-ray diffraction topography (WBXDT) measurements in the reflection mode were carried out at LBNL's ALS Beamline 3.3.2 using an X-ray beam with energy ranging from 4 to 25 keV. Infrared transmission microscopy was also used to study the presence of secondary phases. Figure 5a, b shows an optical photograph of an as-grown $Cd_{0.9}Zn_{0.1}Te_{0.985}Se_{0.015}$ sample with an area of

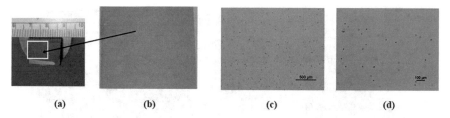

(a)   (b)   (c)   (d)

Fig. 5 (a) Optical photograph and (b) the corresponding X-ray topographic image of the area within the white rectangle (not to the scale) of an as-grown $Cd_{0.9}Zn_{0.1}Te_{0.985}Se_{0.015}$ sample. Sample dimensions: ~1.9 × 1.3 cm^2, (c) and (d) IR transmission microscopic images of the as-grown $Cd_{0.9}Zn_{0.1}Te_{0.985}Se_{0.015}$ sample at two different magnifications. (Taken from Roy et al. [69])

1.9 × 1.3 cm^2. An X-ray topographic image was taken in the area denoted by the white rectangle. As evident from the X-ray topographic image, the sample is free from sub-grain boundaries and their network as opposed to typical CZT samples. The IR transmission microscopic images at two different magnifications for the as-grown $Cd_{0.9}Zn_{0.1}Te_{0.985}Se_{0.015}$ sample are shown in Fig. 5c, d. The presence of high concentrations of Te inclusions is evident from Fig. 5c, d. The concentrations are comparable to conventional CZT with 10% Zn. The Te inclusions in the as-grown $Cd_{0.9}Zn_{0.1}Te_{0.985}Se_{0.015}$ sample were observed to be distributed uniformly [70] over a very large volume of 11 × 10.8 × 19.4 mm^3 as shown in Fig. 6. It is evident that the 1.5 atomic % of selenium in CZTS is sufficient to arrest the formation of sub-grain boundary networks while failing to attain a low concentration of Te inclusions. The CZTS with a composition of 2 atomic % selenium, however, was found to be very effective in reducing the concentration of Te inclusions while maintaining material free from sub-grain boundary networks [69]. The optical photograph and corresponding X-ray topographic image of a typical $Cd_{0.9}Zn_{0.1}Te_{0.98}Se_{0.02}$ sample is shown in Fig. 7a, b. The sample contains two grains that are separated by the grain boundary as shown in Fig. 7a. The X-ray topographic picture (Fig. 7b) depicts the microstructural characteristics of the two corresponding grains. No sub-grain boundary network was observed for both grains as shown in Fig. 7a, b. The presence of very few Te inclusions was observed for the as-grown $Cd_{0.9}Zn_{0.1}Te_{0.98}Se_{0.02}$ samples as shown in Fig. 7c, d. Figure 7c, d shows IR transmission microscopic images at two different magnifications. Similar behavior was observed for CZTS containing 4 atomic % and 7 atomic % of selenium. Although all the samples with different selenium concentrations ranging from 1.5 at. % to 7 at. % was observed to be free from sub-grain boundary networks, the occasional occurrence of an isolated sub-grain boundary was found.

Based on the above observations, 2% of Se in $Cd_{0.9}Zn_{0.1}Te$ was found to be the optimum composition for producing CZTS with highly reduced performance-limiting defects. Since for nuclear detector applications, the charge-transport characteristics are the most important material properties, the optimum composition needs to possess the desired charge-transport properties. Table 1 summarizes the

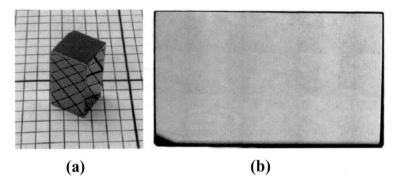

(a)   **(b)**

Fig. 6 (**a**) Optical photograph of a Frisch grid detector sample of dimensions: $11 \times 10.8 \times 19.4$ mm^3 taken from an as-grown $Cd_{0.9}Zn_{0.1}Te_{0.985}Se_{0.015}$ ingot, and (**b**) scanning IR transmission microscopic image of the whole detector. Exposed area: 11×19.4 mm^2. (Taken from Roy et al. [70])

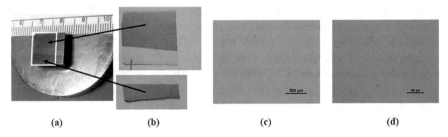

(a)   (b)   (c)   (d)

Fig. 7 (**a**) Optical photograph and (**b**) the corresponding X-ray topographic image of an as-grown $Cd_{0.9}Zn_{0.1}Te_{0.98}Se_{0.02}$ sample. Sample dimensions: ~1.4×1.25 cm^2. The exposed area is shown by the region denoted by the white rectangle (not to the scale); (**c, d**) IR transmission microscopic images of the as-grown $Cd_{0.9}Zn_{0.1}Te_{0.98}Se_{0.02}$ sample at two different magnifications. (Taken from Roy et al. [69])

resistivity and mobility-life time product for electrons [$(\mu\tau)_e$] for all the selenium compositions ranging from 1.5 at. % to 7 at. %. In terms of the charge-transport characteristics, the composition with 2% Se ($Cd_{0.9}Zn_{0.1}Te_{0.98}Se_{0.02}$) was found to be the best. Thus, the composition with 2% Se ($Cd_{0.9}Zn_{0.1}Te_{0.98}Se_{0.02}$) was found to be optimum in terms of the charge-transport characteristics, as well as the microstructure and concentration of Te-rich secondary phases. Roy et al. [69] reported 2 atomic % Se as the optimum composition in terms of charge-transport characteristics and other material properties for THM-grown CZTS samples, while Chaudhuri et al. [63] reported the optimum composition of CZTS compound to be 3 atomic % Se for Bridgman-grown samples. With new quaternary CZTS being recognized as an emerging new material with a high potential to surpass CZT, a more detailed study needs to be performed to find the optimum alloy composition for both Zn and Se to achieve the best detector performance.

Table 1 Resistivity and $(\mu\tau)_e$ values for CZTS with different selenium concentrations

Composition	Resistivity (Ω-cm)	$(\mu\tau)_e$ (cm^2/V)
$Cd_{0.9}Zn_{0.1}Te_{0.93}Se_{0.07}$	$2-5 \times 10^9$	$1-2 \times 10^{-3}$
$Cd_{0.9}Zn_{0.1}Te_{0.96}Se_{0.04}$	$1-1.5 \times 10^{10}$	$2-2.5 \times 10^{-3}$
$Cd_{0.9}Zn_{0.1}Te_{0.98}Se_{0.02}$	$1-3 \times 10^{10}$	6.6×10^{-3} (highest) $4.5-5 \times 10^{-3}$ (average)
$Cd_{0.9}Zn_{0.1}Te_{0.985}Se_{0.015}$	$1-3 \times 10^{10}$	$1-2 \times 10^{-3}$

Taken from Roy et al. [69]

In summary, Se was found to act as an effective element in the CZTS matrix to engineer the concentration of defects and mitigate the long-standing issues pertaining to CZT detector materials. The partial replacement of tellurium by selenium was found to offer multi-faceted advantages. CZTS material was observed to be harder than conventional CZT with 10% Zn, possess less deep trap levels, and have fewer Te-rich secondary phases. Thus, CZTS proved to be superior compared to the presently available CZT in many critical material aspects. The detector performance based on CZTS also found to outperform all other room-temperature semiconductor detectors for many spectroscopic and imaging applications due to the reduction of performance-limiting defects. In consideration of the overall superior material properties and homogeneity of CZTS, the material shows potential to supersede CZT in the future as the leading detector-grade commercial material, particularly as the purity of the Se used in the crystal is further increased.

References

1. Schlesinger, T. E., et al. (2001). Cadmium zinc telluride and its use as a nuclear radiation detector material. *Materials Science and Engineering R, 32*, 103.
2. Yang, G., & James, R. B. (2009). Applications of CdTe, CdZnTe, and CdMnTe radiation detectors. In *Physics, defects, hetero- and nano-structures, crystal growth, surfaces and applications part II, (EDAX, Triboulet R. et al.)* (p. 214). Elsevier.
3. Harrison, F. A., et al. (2013). The nuclear spectroscopic telescope array (NuSTAR) high-energy X-ray mission. *The Astrophysical Journal, 770*, 103.
4. Krawczynski, H. S., et al. (2016). X-ray polarimetry with the polarization spectroscopic telescope array (PolSTAR). *Astroparticle Physics, 75*, 8.
5. Slomka, P. J., et al. (2019). Solid-state detector SPECT myocardial perfusion imaging. *Journal of Nuclear Medicine, 60*, 1194.
6. Triboulet, R. (2005). Fundamentals of the CdTe and CdZnTe bulk growth. *Physica Status Solidi (c), 5*, 1556.
7. Rudolph, P. (1994). Fundamental studies on Bridgman growth of CdTe. *Progress in Crystal Growth and Characterization of Materials, 29*, 275.
8. Jing, W., & Chi, L. (2019). Recent advances in cardiac SPECT instrumentation and imaging methods. *Physics in Medicine and Biology, 64*, 06TR01.
9. Sakamoto, T., et al. (2011). The second swift burst alert telescope gamma-ray burst catalog. *The Astrophysical Journal Supplement Series, 195*, 1.
10. Takahashi, T., et al. (2001). High-resolution CdTe detector and applications to imaging devices. *IEEE Transactions on Nuclear Science, 48*, 287.

11. MacKenzie, J., et al. (2013). Advancements in THM-grown CdZnTe for use as substrates for HgCdTe. *Journal of Electronic Materials, 42*, 3129.
12. Iniewski, K. (2014). CZT detector technology for medical imaging. *Journal of Instrumentation, 9*, 1.
13. Kargar, A., et al. (2011). Charge collection efficiency characterization of a HgI_2 Frisch collar spectrometer with collimated high energy gamma rays. *Nuclear Instruments and Methods in Physics Research A, 652*, 186.
14. Hitomi, K., et al. (2013). TlBr capacitive Frisch grid detectors. *IEEE Transactions on Nuclear Science, 60*, 1156.
15. He, Y., et al. (2018). High spectral resolution of gamma-rays at room temperature by perovskite $CsPbBr_3$ single crystals. *Nature Communications, 9*, 1609.
16. Johns, P. M., & Nino, J. C. (2019). Room temperature semiconductor detectors for nuclear security. *Journal of Applied Physics, 126*, 040902.
17. Takahashi, T., & Watanabe, S. (2001). Recent progress in CdTe and CdZnTe detectors. *IEEE Transactions on Nuclear Science, 48*, 950.
18. Szeles, C., & Eissler, E. E. (1998). Current issues of high-pressure Bridgman growth of semi-insulating CdZnTe. *Materials Research Society Symposium Proceedings, 484*, 309.
19. Szeles, C., et al. (2004). Development of the high-pressure electro-dynamic gradient crystal-growth technology for semi-insulating CdZnTe growth for radiation detector applications. *Journal of Electronic Materials, 33*, 742.
20. Triboulet, R., et al. (1990). "Cold travelling heater method", a novel technique of synthesis, purification and growth of CdTe and ZnTe. *Journal of Crystal Growth, 101*, 216.
21. El Morki, A., et al. (1994). Growth of large, high purity, low cost, uniform CdZnTe crystals by the "cold travelling heater method". *Journal of Crystal Growth, 138*, 168.
22. Shiraki, H., et al. (2009). THM growth and characterization of 100 mm diameter CdTe single crystals. *IEEE Transactions on Nuclear Science, 56*, 1717.
23. Zhang, N., et al. (2011). Anomalous segregation during electrodynamic gradient freeze growth of cadmium zinc telluride. *Journal of Crystal Growth, 325*, 10.
24. Perfeniuk, C., et al. (1992). Measured critical resolved shear stress and calculated temperature and stress fields during growth of CdZnTe. *Journal of Crystal Growth, 119*, 261.
25. Datta, A., et al. (2011). Experimental studies on control of growth interface in MVB grown CdZnTe and its consequences. In *IEEE Nuclear Science Symposium Conference Record 4720.* IEEE.
26. Zhou, B., et al. (2018). Modification of growth interface of CdZnTe crystals in THM process by ACRT. *Journal of Crystal Growth, 483*, 281.
27. Roy, U. N., et al. (2010). Growth and interface study of 2 in diameter CdZnTe by THM technique. *Journal of Crystal Growth, 312*, 2840.
28. Gul, R., et al. (2017). A comparison of point defects in $Cd_{1-x}Zn_xTe_{1-y}Se_y$ crystals grown by Bridgman and traveling heater methods. *Journal of Applied Physics, 121*, 125705.
29. Carini, G. A., et al. (2007). High-resolution X-ray mapping of CdZnTe detectors. *Nuclear Instruments and Methods in Physics Research Section A: Accelerators, Spectrometers, Detectors and Associated Equipment, 579*, 120.
30. Bolotnikov, A. E., et al. (2013). Characterization and evaluation of extended defects in CZT crystals for gamma-ray detectors. *Journal of Crystal Growth, 379*, 46.
31. Yang, G., et al. (2013). Post-growth thermal annealing study of CdZnTe for developing room-temperature X-ray and gamma-ray detectors. *Journal of Crystal Growth, 379*, 16.
32. Bolotnikov, A. E., et al. (2016). CdZnTe position-sensitive drift detectors with thicknesses up to 5 cm. *Applied Physics Letters, 108*, 093504.
33. Szeles, C., et al. (2002). Advances in the crystal growth of semi-insulating CdZnTe for radiation detector applications. *IEEE Transactions on Nuclear Science, 49*, 2535.
34. Triboulet, R. (2003). Crystal growth technology. In H. J. Scheel & T. Fukuda (Eds.), *CdTe and CdZnTe growth* (p. 373). Wiley.
35. Guergouri, K., et al. (1988). Solution hardening and dislocation density reduction in CdTe crystals by Zn addition. *Journal of Crystal Growth, 86*, 61.

36. Imhoff, D., et al. (1991). Zn influence on the plasticity of $Cd_{0.96}Zn_{0.04}Te$. *Journal de Physique III France, 1*, 1841.
37. Buis, C., et al. (2013). Effects of dislocation walls on image quality when using cadmium telluride X-ray detectors. *IEEE Transactions on Nuclear Science, 60*, 199.
38. Johnson, C. J. (1989). Recent Progress in lattice matched substrates for HgCdTe epitaxy. *SPIE, 1106*, 56.
39. Tanaka, A., et al. (1989). Zinc and selenium co-doped CdTe substrates lattice matched to HgCdTe. *Journal of Crystal Growth, 94*, 166.
40. Fiederle, M., et al. (1994). Comparison of CdTe, $Cd_{0.9}Zn_{0.1}Te$ and $CdTe_{0.9}Se_{0.1}$ crystals: Application for γ- and X-ray detectors. *Journal of Crystal Growth, 138*, 529.
41. Roy, U. N., et al. (2014). Growth of $CdTe_xSe_{1-x}$ from a Te-rich solution for applications in radiation detection. *Journal of Crystal Growth, 386*, 43.
42. Roy, U. N., et al. (2015). High compositional homogeneity of $CdTe_xSe_{1-x}$ crystals grown by the Bridgman method. *Applied Physics Letters Materials, 3*, 026102.
43. Roy, U. N., et al. (2014). Evaluation of $CdTe_xSe_{1-x}$ crystals grown from a Te-rich solution. *Journal of Crystal Growth, 389*, 99.
44. Hannachi, L., & Bouarissa, N. (2008). Electronic structure and optical properties of $CdSe_xTe_{1-x}$ mixed crystals. *Superlattices and Microstructures, 44*, 794.
45. Roy, U. N., et al. (2019). Role of selenium addition to CdZnTe matrix for room-temperature radiation detector applications. *Scientific Reports, 9*, 1620.
46. Roy, U. N., et al. (2020). X-ray topographic study of Bridgman-grown CdZnTeSe. *Journal of Crystal Growth, 546*, 125753.
47. Egarievwe, S. U., et al. (2020). Optimizing CdZnTeSe Frisch-grid nuclear detector for gamma-ray spectroscopy. *IEEE Access, 8*, 137530.
48. Hwang, S., et al. (2019). Anomalous Te inclusion size and distribution in CdZnTeSe. *IEEE Transactions on Nuclear Science, 66*, 2329.
49. Roy, U. N., et al. (2019). High-resolution virtual Frisch grid gamma-ray detectors based on as-grown CdZnTeSe with reduced defects. *Applied Physics Letters, 114*, 232107.
50. Kleppinger, J. W., et al. (2021). Growth of $Cd_{0.9}Zn_{0.1}Te_{1-y}Se_y$ single crystals for room-temperature gamma ray detection. *IEEE Transactions on Nuclear Science, 68*, 2429.
51. Roy, U. N., et al. (2019). Evaluation of CdZnTeSe as a high-quality gamma-ray spectroscopic material with better compositional homogeneity and reduced defects. *Scientific Reports, 9*, 7303.
52. Nag, R., et al. (2021). Characterization of vertical Bridgman grown $Cd_{0.9}Zn_{0.1}Te_{0.97}Se_{0.03}$ single crystal for room-temperature radiation detection. *Journal of Materials Science: Materials in Electronics, 32*, 26740.
53. Herraiz, L. M., et al. (2021). Vertical gradient freeze growth of two inches $Cd_{1-x}Zn_xTe_{1-y}Se_y$ ingots with different Se content. *Journal of Crystal Growth, 537*, 126291.
54. Franc, J., et al. (2020). Microhardness study of $Cd_{1-x}Zn_xTe_{1-y}Se_y$ crystals for X-ray and gamma ray detectors. *Materials Today Communications, 24*, 101014.
55. Chang, C. Y., & Tseng, B. H. (1997). Crystal growth of CdTe alloyed with Zn, Se and S. *Materials Science and Engineering: B, 49*(1), 1.
56. Gul, R., et al. (2015). Research update: Point defects in $CdTe_xSe_{1-x}$ crystals grown from a Te-rich solution for applications in detecting radiation. *Applied Physics Letters Materials, 3*, 040702.
57. Yakimov, A., et al. (2019). Growth and characterization of detector-grade CdZnTeSe by horizontal Bridgman technique. *SPIE Proceedings, 11114*, 111141N.
58. Chaudhuri, S. K., et al. (2020). Pulse-shape analysis in $Cd_{0.9}Zn_{0.1}Te_{0.98}Se_{0.02}$ room-temperature radiation detectors. *Applied Physics Letters, 116*, 162107.
59. Chaudhuri, S. K., et al. (2020). Charge transport properties in CdZnTeSe semiconductor room-temperature γ-ray detectors. *Journal of Applied Physics, 127*, 245706.
60. Dědič, V., et al. (2021). Mapping of inhomogeneous quasi-3D electrostatic field in electro-optic materials. *Scientific Reports, 11*, 2154.

61. Pipek, J., et al. (2021). Charge transport and space-charge formation in $Cd_{1-x}Zn_xTe_{1-y}Se_y$ radiation detectors. *Physical Review Applied, 15*, 054058.
62. Park, B., et al. (2022). Bandgap engineering of $Cd_{1-x}Zn_xTe_{1-y}Se_y$ (0< x< 0.27, 0< y< 0.026). *Nuclear Instruments and Methods in Physics Research Section A: Accelerators, Spectrometers, Detectors and Associated Equipment, 1036*, 166836.
63. Chaudhuri, S. K., et al. (2021). Quaternary Semiconductor $Cd_{1-x}Zn_xTe_{1-y}Se_y$ for High-Resolution, Room-Temperature Gamma-Ray Detection. *Crystals, 11*, 7.
64. Franc, J., et al. (2021). Spectral Dependence of the Photoplastic Effect in CdZnTe and CdZnTeSe. *Materials, 14*, 1465.
65. Egarievwe, S. U., et al. (2019). Ammonium fluoride passivation of CdZnTeSe sensors for applications in nuclear detection and medical imaging. *Sensors, 19*, 3217.
66. Rejhon, M., et al. (2018). Influence of deep levels on the electrical transport properties of CdZnTeSe detectors. *Journal of Applied Physics, 124*, 235702.
67. Guskov, V. N., et al. (2004). Vapour pressure investigation of CdZnTe. *Journal of Alloys and Compounds, 371*, 118.
68. Brill, G., et al. (2005). Molecular beam epitaxial growth and characterization of Cd-based II-VI wide-bandgap compounds on Si substrates. *Journal of Electronic Materials, 34*, 655.
69. Roy, U. N., et al. (2021). Optimization of selenium in CdZnTeSe quaternary compound for radiation detector applications. *Applied Physics Letters, 118*, 152101.
70. Roy, U. N., et al. (2019). Characterization of large-volume Frisch grid detector fabricated from as-grown CdZnTeSe. *Applied Physics Letters, 115*, 242102.

Investigation of Charge Transport Properties and the Role of Point Defects in CdZnTeSe Room Temperature Radiation Detectors

Sandeep K. Chaudhuri, Ritwik Nag, Joshua W. Kleppinger, and Krishna C. Mandal

1 Introduction

Extensive research in the past three decades on high-atomic number (high-Z)-based semiconductor resulted in the now-matured wide bandgap ternary compound semiconductor $Cd_{1-x}Zn_xTe$ (CZT) for gamma-photon detection at room temperature [1–10]. Before the advent of high-resolution CZT detectors, the low-temperature requirement for operation of existing high-resolution HPGe (high-purity germanium) detectors for high-energy gamma ray detection was strict enough to resort to lower-resolution scintillation detectors or indirect readout detection system especially for imaging systems. Resolution and contrast of radiological images nevertheless depend directly on the energy resolution of the detectors used in the imaging system. Low-resolution and poor-contrast images may lead to both over- and underdiagnosis, which collectively is one of the major impediments in gaining public confidence in applying radiography for screening deadly diseases such as breast cancer [11]. Indirect conversion imaging system such as those comprising scintillator detectors relies heavily on detecting photons through several photomultipliers. The algebraic sum of signals from all the photomultiplier tubes constitutes the energy signal, while the weighted sum of the signals gives the spatial location of the event in the scintillator array. In this process, if a photon is lost due to self-absorption within the crystal or at the coupling window (at the scintillator-PMT interface), it may lead to miscalculations in the energy of the photons and the location of the interactions [12].

Semiconductor detectors such as CZT are direct readout devices where a photon interacts with a detector to produce charge pairs (electrons and holes) which are

S. K. Chaudhuri · R. Nag · J. W. Kleppinger · K. C. Mandal (✉)
Department of Electrical Engineering, University of South Carolina, Columbia, SC, USA
e-mail: mandalk@cec.sc.edu

collected by one of the many pixels (anode) and the cathode [13]. The drifting charge carriers constitute an electric current which is directly converted into an electrically measurable signal by the front-end electronics. Such semiconductor detectors, apart from being direct converter, have other major advantages as well. Because of their high material density, they can absorb high-energy neutral particles such as gamma photons with a high efficiency yet within a compact volume. Moreover, high-Z constituents such as those in CZT enhance the detection efficiency and resolution further as the interaction probability of gamma photons with matter increases manifold through photoelectric absorption – a mechanism where a photon deposits its total energy in a single interaction event [14]. Further, the bandgap of compound semiconductors can be engineered by varying the stoichiometry to optimize the absorption efficiency and the leakage current. For instance, inclusion of 10% Zn in CdTe ($Cd_{0.9}Zn_{0.1}Te$) can increase the bandgap by 4% [15]. Leakage current of the detectors controls the electronic noise and hence the detector resolution. Wide bandgap results in high electrical resistivity ($\geq 10^{10}$ Ω-cm) and hence low leakage currents (~1 nA or less) even for thin (1–2 mm) detectors. Also, the charge transport properties for at least one type of carriers are high enough to obtain fast and efficient charge collection resulting in high timing, spatial, and energy resolutions. The above solid-state advantages coupled to the advancement of digital signal processing such as tomography and signal correction techniques have elevated the present-day radiological imaging to a state where image qualities are obtained with unprecedented details [12]. Unfortunately, the success story of CZT is limited due to the unavailability of large volume detector-grade single crystals due to its inherent poor crystal growth and crystal defects [16–23]. Large-volume CZT detectors invariably need either special electrode geometry to obtain single polarity charge transport or signal correction methods [13, 24–26]. As a result, the cost of production of CZT detectors increases limiting the commercial viability of CZT-based medical imaging systems.

$Cd_{1-x}Zn_xTe_{1-y}Se_y$ (CZTS) is a quaternary successor of ternary compound semi-conductor CZT which has been discovered recently and has been demonstrated to provide a crystal growth yield higher than 90% [18, 27–31]. The addition of selenium in small amount (2–3 at. %) in the CZT matrix has been reported to reduce the lattice stress. Lattice strain in CZT arises due to the Zn-segregation-related compositional nonuniformity throughout the ingot and due to the high concentration of sub-grain boundary network because of the poor thermophysical properties of the CZT melt [32]. Addition of Se in the CZT matrix has also been reported to reduce the formation of Cd vacancies [33]. The performance of the CZTS as low-cost room temperature gamma detector has already equaled that of the CZT detectors [34, 35]. Some of the physical properties of CZTS relevant to radiation detection have been compared to those of CZT in Table 1. The energy resolution of the present-day CZTS detectors is limited by the charge trapping, especially hole trapping, in point defects [36–42]. To improve the performance of the CZTS detector further and make it a future-generation room-temperature detector material, it is of utmost importance to characterize and study the electrically active point defects responsible for charge trapping.

Table 1 Comparison of physical properties relevant to radiation detection between CZT and CZTS

Detector Material	Bandgap (eV)	Resistivity (Ω-cm)	$\mu_e$cm^2/V.s	$\mu_h$cm^2/V.s	$\mu_e\tau_e$cm^2/V	$\mu_h\tau_h$cm^2/V	Energy resolution for 662 keV γ-rays
CZT	1.58	5×10^{10}	1000	120	7.0×10^{-3}	1.2×10^{-4}	0.48%[a]
CZTS	1.52	3×10^{10}	964	55	6.5×10^{-3}	1.4×10^{-4}	0.77%[b]

[a]Pixelated detectors with correction
[b]Frisch grid detector without correction

The present chapter discusses the growth of single crystalline CZTS detectors using a modified vertical Bridgman (VBM) and a vertical gradient freeze (VGF) growth method and their characterization using various electrical and spectroscopic techniques. The formation of defects in the CZTS crystals studied using ab initio calculations has been correlated with the defects observed through photo-induced current transient spectroscopy (PICTS).

2 Growth of CZTS Single Crystals

CZTS single crystals have been grown using a modified vertical Bridgman method and a gradient freeze method. For both the methods, 99.999% (5 N) pure elemental precursors (Cd, Zn, Te, and Se) have been zone refined (ZR) in-house to obtain 7 N purity materials. The ZR precursors are weighed in stoichiometric ratio to obtain the compound $Cd_{0.9}Zn_{0.1}Te_{0.97}Se_{0.03}$. An excess of 5% tellurium (Te) was added to reduce the growth temperature [43]. A 15-ppm compensating indium (In) of 7 N purity was used as a dopant to obtain the desired resistivity [44]. The precursors along with In has been sealed in a conically tipped quartz ampoule under a vacuum of 10^{-6} Torr. The quartz tube has been carbon coated to prevent Cd from reacting with the inner surface of the quartz tube [45]. The quartz tubes used for the growths have an inner diameter of ≥ 16 mm with a wall thickness of ≥ 2 mm. The quartz ampoule with precursor materials has been fused with a quartz tie rod and suspended inside the furnace.

2.1 Vertical Bridgman Method

For the vertical Bridgman method growth, a three-zone Lindberg blue furnace has been used. The three heaters have been programmed separately to establish the growth profile and to maintain an axial temperature gradient of 3.5 °C/cm. Figure 1a shows the temperature profile for the VBM growth which comprises 3 hrs heat treatment at three progressively higher temperatures of 890, 970, and 980 °C to synthesize the polycrystalline charges from the precursors. The quartz rod has been attached to a movable frame controlled by a DC motor to lower the quartz tube into the hot zone at a constant velocity of 3 mm/hr – a standardized speed to solidify the melt directionally in a crystalline form [46]. The tie rod has been attached to a separate motor to rotate the ampoule about its axis at a speed of 15 rpm throughout the growth period to obtain a uniform radial temperature distribution. The furnace has been cooled slowly at various ramp rates following the growth cycle to avoid any thermal stress. Figure 1b shows the photograph of a VBM-grown CZTS ingot.

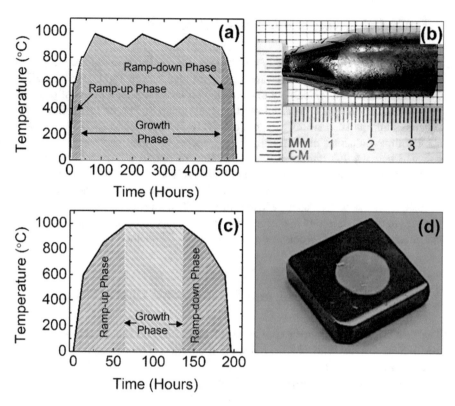

Fig. 1 (**a**) Temperature profile for the VBM growth of CZTS crystals. (**b**) A VBM-grown $Cd_{0.9}Zn_{0.1}Te_{0.97}Se_{0.03}$ crystal ingot. (**c**) Temperature profile for the VGF growth of CZTS crystals. (**d**) A photograph of a planar CZTS detector grown using a VGF method

2.2   Vertical Gradient Freeze Method

A vertical gradient freeze or VGF method has many advantages over the VBM or THM growth. VGF method not only allows to achieve a relatively higher growth rate; it also minimizes thermal drift or temperature fluctuations as there does not exist any relative translational motion between the ampoule and the heater. Thermal drift or temperature fluctuations lead to uncontrolled changes in the growth interface [47]. A furnace-ampoule arrangement similar to the VBM growth has been adopted except for the vertical translational movement. The upper and the middle heating zones out of the three zones have been programmed to set the desired temperature profile. The top hot zone and the middle cold zone are situated 20 inches apart, and the ampoule has been placed within the top zone with the tip at the level of the cold zone. The top heater has been ramped up to 985 °C and the middle heater to 960 °C. After a synthesis duration of 72 hrs, both the heating zones were ramped down at very slow rates of 5–10 °C/hr. The temperature growth profile for the VGF growth has been summarized in Fig. 1c.

2.3 Crystal Cutting and Detector Fabrication

The quartz ampoules are carefully cut using a diamond saw to release the grown ingot. The ingots are sliced into wafers using a Extec Labcut 150 wafer cutter. The wafers are further cut typically with a rectangular cross section for ease of processing. The wafers are hand polished mechanically using SiC paper first followed by aqueous alumina suspension with progressively finer grit sizes (5–0.02 μm) and a microfiber polishing cloth. The polished wafers are then chemo-mechanically polished using a 2% bromo-methanol solution to obtain a mirror like finish on all the surfaces. The polished wafers are thoroughly cleaned with isopropanol and deionized water prior to metal contact deposition. Identical gold contacts have been deposited on the wafers using a Quorum Q150T sputter coater on two opposite surfaces to obtain detectors in planar configuration. Figure 1d shows the photograph of a finished planar CZTS detector grown using a VGF method.

3 Structural, Electrical, and Spectroscopic Characterization

The grown crystal has been characterized to investigate the crystallinity, composition, resistivity, and charge transport properties. The crystallinity has been investigated by powder x-ray diffraction, and the compositional analysis has been carried out using energy-dispersive analysis by x-rays (EDAX) spectroscopy. The resistivities and leakage currents have been determined from the current-voltage characteristics, while the charge transport properties have been measured using alpha-ray spectroscopic methods.

3.1 Structural and Compositional Characterization

Figure 2a shows an x-ray powder diffractogram (XRD) of the VBM-grown $Cd_{0.9}Zn_{0.1}Te_{0.97}Se_{0.03}$ crystal. The diffractogram has been collected using a Rigaku Ultima IV D/Max 2100 powder diffractometer using CuK_α radiation ($\lambda = 1.5418$ Å). Sharp and strong diffraction peaks at $2\theta = 23.9°$, $39.5°$, and $46.5°$ along with relatively lower intensity peaks at $2\theta = 57.1°$, $62.8°$, $71.7°$, and $76.8°$ have been observed in the scan range $10° \leq 2\theta \leq 80°$. The crystal planes corresponding to the XRD peaks have been identified by comparing the peak positions with that of zincblende $Cd_{0.9}Zn_{0.1}Te$ (ICSD# 620556) [46]. The strongest peak has been identified as the (111) plane indicating that the prevalent growth direction is perpendicular to the (111) plane. The narrow distribution of the peaks indicates toward the high degree of crystallinity. Similar XRD results have been observed for the VGF-grown CZTS.

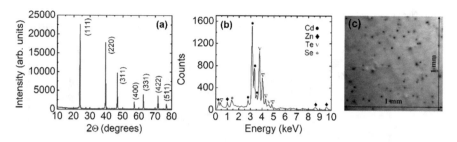

Fig. 2 (**a**) An x-ray powder diffractogram of the VBM grown $Cd_{0.9}Zn_{0.1}Te_{0.97}Se_{0.03}$ crystal. (**b**). Energy-dispersive x-ray spectrum obtained from a representative part of the VGF-grown $Cd_{0.9}Zn_{0.1}Te_{0.97}Se_{0.03}$ ingot. (**c**) IR transmission image of a 1 mm × 1 mm portion of a VGF-grown CZTS crystal

Figure 2b shows one of the several energy-dispersive analysis by x-ray (EDAX) spectra obtained from different representative parts of a VGF-grown $Cd_{0.9}Zn_{0.1}Te_{0.97}Se_{0.03}$ ingot. The EDAX spectra in the present samples have been acquired using a Tescan Vega 3 SBU variable pressure high-resolution scanning electron microscope (SEM) equipped with EDAX microanalysis software. Considering that the peak intensity corresponding to a certain element in the EDAX spectrum is proportional to the atomic percentage of that element present in the sample, the Se-to-Te atomic ratio has been calculated from the EDAX spectra to be 0.03 ± 0.02 which is the intended stoichiometry. The Zn-to-Cd atomic ratio, however, was found to be 0.08 ± 0.02 which is little less than the intended value of 0.1. The Se-to-Te and the Zn-to-Cd atomic ratio have been calculated to be 0.04 ± 0.02 and 0.1 ± 0.02, respectively, for the VBM-grown CZTS. The atomic percentage of the elemental composition is thus found to be close to the intended stoichiometry within the experimental error bar.

Figure 2c shows the infrared (IR) transmission image of a 1 mm × 1 mm portion of a VGF grown CZTS crystal using a IR transmission microscopy system comprising a large field-of-view microscope objective, a CCD camera, a motorized X-Y-Z translation stages, and a light source coupled with a wide-beam condenser for illuminating the samples [48]. IR transmission image provides direct information on the Te secondary phase formation which unlike CZTS is not transparent to IR light. The contrast image indicates the presence of Te inclusions with size above 1 μm. The IR transmission image obtained for the VGF-grown CZTS samples indicate the presence of Te inclusions with average size <9 μm. Te inclusions with size above 10 μm is known to trap charges [48]. The properties of the Te inclusions in the VBM grown crystal were quite similar.

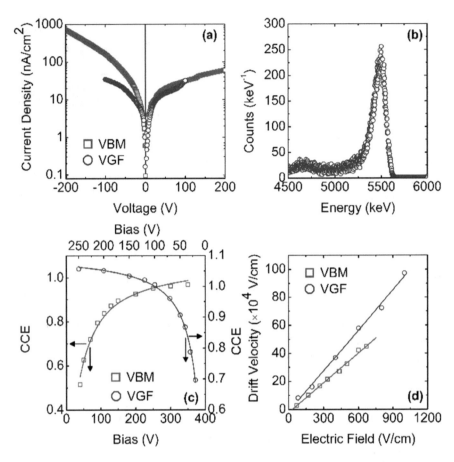

Fig. 3 (**a**) Current density-voltage (*J-V*) characteristics obtained for a VBM- and VGF-grown CZTS detector measured at room temperature. (**b**) An optimized pulse height spectrum acquired using a VGF-grown CZTS detector exposed to a Am-241 alpha emitting source. (**c**) Variation of charge collection efficiency as a function of the bias voltage for a VGF- and a VBM-grown detector. Solid lines are the single carrier Hecht equation fit. (**d**) Variation of electron drift velocity as a function of the electric field obtained for VGF- and VBM-grown CZTS detectors. The solid lines show the linear fits for the electron drift mobility determination

3.2 Electrical and Spectroscopic Characterization

The wafers and the polished crystals/detectors with the intended compositional and structural characteristics have been selected for CZTS crystal property investigation. The electrical bulk resistivity has been measured through current-voltage scans, and the charge transport properties have been measured through alpha particle spectroscopy.

Figure 3a shows the current density-voltage (*J-V*) characteristics obtained for the VBM- as well as the VGF-grown detector measured at room-temperature using

Table 2 Detector parameters obtained from electrical and spectroscopic characterizations

Detector	Sizes (mm^3)	Contact area (cm^2)	Resistivity (Ω-cm)	Drift-mobility (cm^2/V.s)	$\mu_e \tau_e$(cm^2/V)
VBM grown	4.0 × 4.0 × 1.7	0.07	1 × 10^{10}	710	1.5 × 10^{-3}
VGF grown	11.0 × 11.0 × 3.0	0.07	3 × 10^{10}	964	3 × 10^{-3}

a Keithley 237 source-measure unit. The detector and the contact dimensions have been given in Table 2. The *J-V* characteristics for the VGF detector were found to be relatively more symmetric with respect to the bias polarity than that of the VBM detector. Such observed asymmetry although is likely to be linked to the surface properties rather than bulk properties. Variation of surface properties (such as dangling bonds, surface, and interface states) between the two opposite surfaces chosen for contact deposition leads to different surface carrier recombination velocities causing the current to vary with respect to bias polarity [49]. The leakage current density of both the detectors was found to be ≈30 nA/cm^2 at +100 V. The resistivity of the VGF-grown detector, calculated from a lower range (-1 to $+1$ V) *J-V* characteristic, was found to be 3 × 10^{10} Ω-cm – higher by a factor of 3 than that found for the VBM-grown detector. It is to be further investigated if the observed difference in the bulk resistivity is due to the different growth method or surface preparation.

The mobility-lifetime product has been calculated from analog pulse height spectroscopy using an alpha particle source. The spectrometer used to acquire the analog pulse height spectrum (PHS) is a standard benchtop alpha spectrometer comprising a Cremat CR110 charge-sensitive preamplifier, an Ortec 672 spectroscopy amplifier, and a Canberra Multiport II multichannel analyzer [20, 50]. The detector has been placed in a shielded test box with a 0.9 μCi ^{241}Am radioisotope emitting 5486 keV (E_i) alpha particles. The test box has been continually evacuated during the measurements. A typical pulse height spectrum acquired using a VGF-grown CZTS detector is shown in Fig. 3b. Figure 3c shows the variation of the charge collection efficiency (CCE, η) as a function of the bias voltage for a VGF- and a VBM-grown detector. The CCE has been defined as the fraction of injected charges that is collected by the spectrometer and is measured as the ratio $\eta = E_p/E_i$, where E_i is the energy of the ionizing radiation emitted by the radiation source and E_p is the energy registered by a calibrated spectrometer from the PHS. As the penetration depth of 5486 keV alpha particles in CZTS is small (<20 μm) and the contact facing the source is negatively biased, the radiation-induced signals are primarily caused by the electron transit. The solid lines in Fig. 3c are the single polarity (electron) Hecht equation (Eq. 1 below) fit to the variation of CCE vs bias voltage. The mobility-lifetime ($\mu\tau$) products are extracted from the Hecht equation fit and are found to be 3 × 10^{-3} cm^2/V and 1.5 × 10^{-3} cm^2/V for the VGF- and the VBM-grown crystals, respectively [28, 51, 52].

$$\eta = \frac{\mu \tau V}{d^2} \left[1 - \exp\left(\frac{-d^2}{\mu \tau V}\right) \right] \tag{1}$$

d being the detector thickness and V the applied bias.

The carrier drift mobility μ, defined as the drift velocity (v_d) acquired per unit applied electric field, is another important charge transport property to be studied. A time-of-flight alpha spectroscopic method has been adopted to measure the drift velocities as a function of bias voltage using a digital alpha particle spectrometer [28, 52]. The digital spectrometer comprises the same test box and the charge sensitive preamplifier as used in the analog spectrometer. The output of the preamplifier was fed to a fast NI5122 digitizer controlled by a custom-built LabVIEW data acquisition program to digitize and store the preamplifier charge pulses. The program also determines the risetime of each recorded pulse and calculates the drift velocity assuming the electric field (E) is linear along the drift direction and uniformly distributed laterally, and only one polarity charge carrier (electron in the present case) transits the full detector thickness at a set bias. More details on the TOF measurement and the related setup could be found elsewhere [20, 28]. Figure 3d shows the plots of electron drift velocity as a function of the electric field. Linear fits of the plots yield the electron mobility as the drift velocity and the electric field are related as $v_d = \mu E$. The electron drift velocities were calculated to be 964 cm^2/V.s and 710 cm^2/V.s for the VGF- and the VBM-grown detectors, respectively. Table 2 summarizes all the device properties discussed so far.

4 Density Functional Theory Calculations: Defect Formation

The indications that addition of Se in the CZT matrix improves the compositional homogeneity need thorough investigation involving both theoretical calculations and direct experimental studies. In this chapter we focus on the probability of formation of various point defects that participate in charge trapping. Density functional theory (DFT)-based calculations were performed to determine the formation energies of vacancy and antisite defects. The formation energy E_{form}, defined as the excess energy needed by a defective supercell to evolve from a defect-free supercell, decides the likelihood of formation of a defect. The lower the formation energy is, the higher is the probability of its appearance. The formation energy which is a function of the defect charge q and the Fermi energy E_f is expressed as

$$E_{form}(E_f, q) = E(q) - E_{bulk} - \sum_i n_i \left(\mu_i + E_i\right) + q\left(E_f + E_v\right) + E_{corr} \tag{2}$$

where $E(q)$ is the energy of the defective cell, n_i is the number of atoms of species i added or removed, μ_i is the chemical potential of atoms, E_f is the Fermi energy, E_v

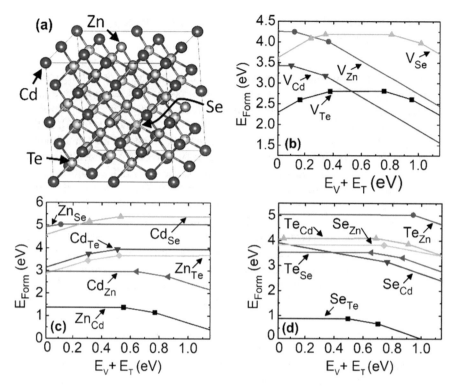

Fig. 4 (**a**) A primitive CZTS supercell where one Te atom replaced by a Se atom. (**b–d**) Formation energy plots for various vacancy and antisite defects in CZTS as a function of their energy locations

is the valence band edge energy, and E_{corr} is a correction factor for periodic imaging [53, 54].

In the present calculation, a Vienna ab initio simulation package (VASP) has been used to implement a generalized gradient approximation (GGA) with PBE pseudopotentials for calculating the formation energy of a 64-atom supercell with a Zn and Se atomic percentage of 12.5% and 3.125% in the supercell [55]. The CZTS supercell has been created by randomly replacing some of the Te atoms from their sites by Se atoms. Figure 4a shows a primitive CZTS supercell where one Te atom has been replaced by a Se atom and Fig. 4b–d shows the formation energy plots for various vacancy and antisite defects. It has been found that the formation energies of Te_{Zn} and Te_{Cd} antisites are at least 1 eV higher than that calculated in a similar CZT supercell which indicates that the formation probabilities of these defects are less probable in CZTS compared to that in CZT. It is worth noting that the Te_{Cd}^{++}, a doubly charged Te antisite, is a deep trap level situated at 0.78 eV below the conduction band minimum and is one of the most potential intrinsic trap centers observed in CZT [37]. The DFT calculations thus show that addition of Se in the CZT matrix does help in reducing the formation of potential charge trapping centers.

5   Photo-Induced Current Transient Spectroscopy Measurements

Thermally stimulated transient measurement techniques such as deep-level transient spectroscopy (DLTS), current transient spectroscopy (CTS), and charge transient spectroscopy (QTS) are extremely sensitive tools to investigate deep-level trap centers or point defects in junction-based semiconductor devices [56–60]. In these techniques, the junction-based device (p-n or Schottky diode) is kept at a quiescent (steady-state) reverse bias V_r. An electric pulse (filling pulse) momentarily drives the junction to a forward bias during which the depletion width collapses, and the empty trap states are filled with the majority carriers. On the termination of the filling pulse, the traps are reemitted (detrapped) to the energetically favorable band with an emission rate dependent on the device temperature. Device properties such as junction capacitance, current, and trapped electric charge change as a function of time during the detrapping resulting in capacitance, current, or charge transients.

In the case of high resistivity semiconductors such as CZTS, forming a rectifying junction is quite difficult. In such samples, the trap filling is done optically by illuminating the device with photons of appropriate energy (greater than the bandgap of the semiconductor) for a duration long enough to ensure a saturation trap filling [37]. These devices often have very small capacitances which demands measurement of current transients instead of capacitance. The technique is therefore called photo-induced current transient spectroscopy. Figure 5a shows the schematic of the PICTS setup used in this study. The PICTS system comprises a 635 nm laser powered by a function generator (GW Instek AFG-2105) to obtain an excitation pulse (Fig. 5b) 2 msec wide which repeats with a frequency of 22 Hz.

The transients formed after the termination of the optical pulse are usually exponential in nature and are function of the emission rates (e_n) and other parameters. An expression for the transients obtained for electron trap centers is given in Eq. 3 [57].

$$i(t) = \frac{eA\mu_n\tau_n V_r N_T}{d} e_n \exp\left(-e_n t\right) \tag{3}$$

where e is the elemental charge of an electron, A is the active contact area, $\mu_n\tau_n$ is the electron mobility-lifetime product, N_T is the trap concentration, d is the detector thickness, and e_n is the emission rate. Figure 5c shows a typical PICTS signal obtained for the present detector at room temperature.

The difference in the current amplitudes $\Delta i = i(t_1) - i(t_2)$ in the transients at two given times t_1 and t_2 ($t_1 - t_2 = \tau$, also called the *rate window*) forms the PICTS signal. Temperature scans are performed to obtain a variation of the signals with temperature. Peaks are formed at temperatures (T_m) where the transient signal amplitude becomes maximum – a resonant condition where $\tau^{-1} = e_n$. Traps located at different energy levels (E_T) within the bandgap form peak at different temperatures. The temperature scans are repeated with different rate window

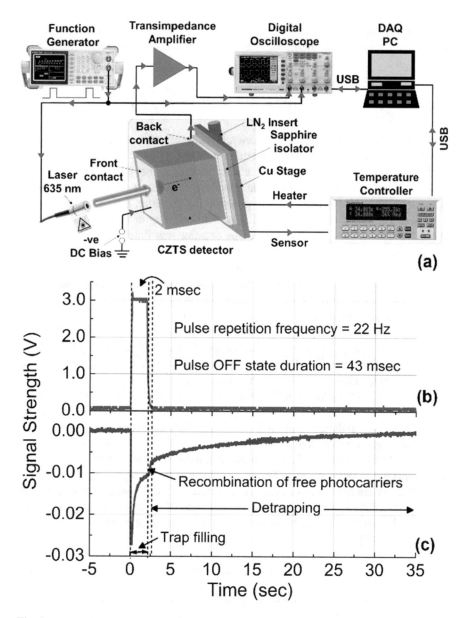

Fig. 5 (**a**) A schematic of the PICTS setup used to study point defects in CZTS detectors. (**b**) An excitation pulse used for saturation trap filling. (**c**) A typical PICTS signal obtained at room temperature for the VBM-grown CZTS detector

settings. Different rate window settings basically change the resonant condition thereby shifting the peaks for each trap center slightly on the temperature scale. The correlation of the emission rates and the peak position (T_m) for each trap center are given by Eq. 4 from which the Arrhenius plots ($\ln e_n/T^2$ vs. $1/T$) are obtained to determine the activation energy $E_a = E_c - E_T$ and capture cross sections σ_n of the traps.

$$e_n = \sigma_n \langle v_{th,n} \rangle N_c T^2 \exp\left(-\frac{E_c - E_T}{k_B T}\right) \tag{4}$$

where $\langle v_{th,n} \rangle$ is the mean thermal velocity, N_c is the effective density of states of the conduction band, E_c conduction band minimum, and k_B is Boltzmann's constant.

Figure 6a shows a PICTS spectrum recorded for the VBM-grown Cd$_{0.9}$Zn$_{0.1}$Te$_{0.98}$Se$_{0.03}$ detector in the temperature scan range 85–400 K. A digital two-gate algorithm has been used to generate the spectrum for an initial delay of 12 msec. The initial delay is the timing of the beginning of the rate window after the extinction of the excitation pulse. The detector was illuminated by the pulsed laser on the semitransparent electrode biased at a quiescent voltage $V_r = -12$ V. The PICTS spectrum shows the presence of at least three peaks labeled as P1, P2, and P3. Figure 6b shows the Arrhenius plot corresponding to the three peaks obtained by using up to 12 rate windows. The activation energies calculated for the traps corresponding to P1, P2, and P3 are 0.08 ± 0.01 eV, 0.32 ± 0.02 eV, and 0.72 ± 0.07 eV, respectively. Since the electrode illuminated with the laser was biased at a negative potential, the current transient signals were generated by electron flow, and the PICTS signals are predominantly due to electron trapping and detrapping. Hence the activation energies correspond to the energy location of the electron trap centers within the bandgap measured from the conduction band minimum. The defect level corresponding to peak P1 is a shallow one and possibly is not a potential lifetime killer defect. The activation energy of the trap center corresponding to P2 does not match with any of the point defects determined from the DFT calculations. It may be due to a more complex defect structure such as complexes of point defects. The trap center corresponding to the peak P3 has an activation energy close to 0.78 eV reported for Te$_{Cd}^{++}$ electron trap levels in THM-grown CZTS [37]. The volume concentration of the trap centers, N_t, has been calculated using Eq. 5, which assumes that the $\mu\tau$ product remains constant with temperature [57].

$$\Delta i\,(T_m) = q.d.V_r.\mu.\tau.N_t/2.718.t_1 \tag{5}$$

where d is the thickness of the layer in which the traps are saturated and has been assumed to be equal to the detector thickness. The trap concentration calculated from the Eq. 5 is on the order of 10^{11} cm^{-3} which is one order of magnitude lower compared to the trap concentrations observed by Cavallini et al., in VBM-grown detector grade CZT crystals through PICTS measurements [16]. The VGF-grown crystal did not show any resolvable peaks in the PICTS scans which may indicate a defect concentration lower than the sensitivity of the PICTS setup.

Fig. 6 (a) A PICTS spectrum recorded for a VBM grown $Cd_{0.9}Zn_{0.1}Te_{0.98}Se_{0.03}$ detector in the temperature scan range 85–400 K. (b) The Arrhenius plot corresponding to the peaks PI, P2, and P3 observed in the PICTS scans

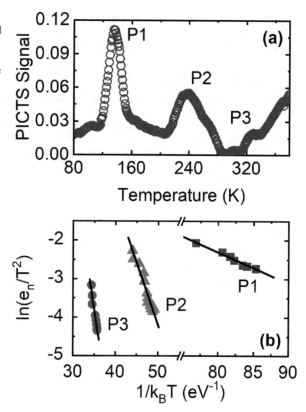

Acknowledgments The authors acknowledge the financial support provided by the DOE Office of Nuclear Energy's Nuclear Energy University Programs (NEUP), Grant No. DE-NE0008662. The work was also partially supported by the Advanced Support Program for Innovative Research Excellence-I (ASPIRE-I) of the University of South Carolina (UofSC), Columbia, Grant No. 15530-E419 and 155312 N1600 and by Los Alamos National Laboratory/DOE (Grant No. 143479).

References

1. Schlesinger, T. E., Toney, J. E., Yoon, H., Lee, E. Y., Brunett, B. A., Franks, L., & James, R. B. (2001). Cadmium zinc telluride and its use as a nuclear radiation detector material. *Materials Science and Engineering, 32*, 103–189.
2. Szeles, C. (2004). CdZnTe and CdTe materials for X-ray and gamma ray radiation detector applications. *Physica Status Soldi B, 241*, 783–790.
3. McGregor, D. S., He, Z., Seifert, H. A., Rojeski, R. A., & Wehe, D. K. (1998). CdZnTe semiconductor parallel strip Frisch grid radiation detectors. *IEEE Transactions on Nuclear Science, 45*, 443–449.

4. Bell, S. J., Baker, M. A., Duarte, D. D., Schneider, A., Seller, P., Sellin, P. J., Veale, M. C., & Wilson, M. D. (2017). Performance comparison of small-pixel CdZnTe radiation detectors with gold contacts formed by sputter and electroless deposition. *Journal of Instrumentation, 12*, 06015.

5. Berrett, H. H., Eskin, J. D., & Barber, H. B. (1995). Charge transport in arrays of semiconductor gamma-ray detectors. *Physical Review Letters, 75*, 156–159.

6. Luke, P. N. (1998). Single-polarity charge sensing in ionization detectors using coplanar electrodes. *Applied Physics Letters, 65*, 2884–2886.

7. Del Sordo, S., Abbene, L., Caroli, E., Mancini, A. M., Zappettini, A., & Ubertini, P. (2009). Progress in the development of CdTe and CdZnTe semiconductor radiation detectors for astrophysical and medical applications. *Sensors, 9*, 3491–3526.

8. Chaudhuri, S. K., & Mandal, K. C.(2022). In: Iniewski K. (ed.) Advanced materials for radiation detection, pp. 211–234. Springer, Cham.

9. Abbene, L., Gerardi, G., Principato, F., Buttacavoli, A., Altieri, S., Protti, N., Tomarchio, E., Del Sordo, S., Auricchio, N., Bettelli, M., Amadè, N. S., Zanettini, S., & Zappettini, A. (2020). Recent advances in the development of high-resolution 3D cadmium–zinc–telluride drift strip detectors. *Journal of Synchrotron Radiation, 27*, 1564–1576.

10. Abbene, L., Principato, F., Gerardi, G., Buttacavoli, A., Cascio, D., Battelli, M., Amadè, N. S., Seller, P., Veale, M. C., Fox, O., Sawhney, K., Zanettini, S., Tomarchio, E., & Zappettini, A. (2020). Room-temperature X-ray response of cadmium–zinc–telluride pixel detectors grown by the vertical Bridgman technique. *Journal of Synchrotron Radiation, 27*, 319–328.

11. Esserman, L. (2020). We need more evidence to answer questions about screening. *Nature, 579*, S5.

12. Iniewski, K. (2014). CZT detector technology for medical imaging. *Journal of Instrumentation, 9*, C11001.

13. Chaudhuri, S. K., Nguyen, K., Pak, R. O., Matei, L., Buliga, V., Groza, M., Burger, A., & Mandal, K. C. (2014). Large area $Cd_{0.9}Zn_{0.1}Te$ pixelated detector: Fabrication and characterization. *IEEE Transactions on Nuclear Science, 61*, 793–798.

14. Knoll, G. F. (2000). *Radiation detection and measurements* (3rd ed.). Wiley.

15. Fougeres, P., Siffert, P., Hageali, M., Koebel, J. M., & Regal, R. (1999). CdTe and $Cd_{1-x}Zn_xTe$ for nuclear detectors: Facts and fictions. *Nuclear Instruments and Methods in Physics Research A, 428*, 38–44.

16. Cavallini, A., Fraboni, B., Castaldini, A., Marchini, L., Zambelli, N., Benassi, G., & Zappettini, A. (2013). Defect characterization in fully encapsulated CdZnTe. *IEEE Transactions on Nuclear Science, 60*, 2870–2871.

17. Bolotnikov, A. E., Camarda, G. S., Carini, G. A., Cui, Y., Li, L., & James, R. B. (2007). Cumulative effects of Te precipitates in CdZnTe radiation detectors. *Nuclear Instruments and Methods in Physics Research A, 571*, 687–698.

18. Roy, U. N., Camarda, G. S., Cui, Y., Gul, R., Yang, G., Zazvorka, J., Dedic, V., Franc, J., & James, R. B. (2019). Evaluation of CdZnTeSe as a high-quality gamma-ray spectroscopic material with better compositional homogeneity and reduced defects. *Scientific Reports, 9*, 7303.

19. Mandal, K. C., Kang, S. H., Choi, M., Bello, J., Zheng, L., Zhang, H., Groza, M., Roy, U. N., Burger, A., Jellison, G. E., Holcomb, D. E., Wright, G. W., & Williams, J. A. (2006). Simulation, modeling, and crystal growth of $Cd_{0.9}Zn_{0.1}Te$ for nuclear spectrometers. *Journal of Electronic Materials, 35*, 1251–1256.

20. Mandal, K. C., Kang, S. H., Choi, M., Kargar, A., Harrison, M. J., McGregor, D. S., Bolotnikov, A. E., Carini, G. A., Camarda, G. C., & James, R. B. (2007). Characterization of low-defect $Cd_{0.9}Zn_{0.1}Te$ and CdTe crystals for high-performance Frisch collar detectors. *IEEE Transactions on Nuclear Science, 54*, 802–806.

21. Mandal, K. C., Krishna, R. M., Muzykov, P. G., & Hayes, T. C. (2012). Fabrication and characterization of high barrier $Cd_{0.9}Zn_{0.1}Te$ Schottky diodes for high resolution nuclear radiation detectors. *IEEE Transactions on Nuclear Science, 59*, 1504–1509.

22. Krishna, R. M., Muzykov, P. G., & Mandal, K. C. (2013). Electron beam induced current imaging of dislocations in $Cd_{0.9}Zn_{0.1}Te$ crystal. *Journal of Physics and Chemistry of Solids, 74*, 170–173.
23. Pak, R. O., & Mandal, K. C. (2015). Defect levels in nuclear detector grade $Cd_{0.9}Zn_{0.1}Te$ crystals. *ECS Journal of Solid State Science and Technology, 5*, P3037–P3040.
24. Chaudhuri, S. K., Krishna, R. M., Zavalla, K. J., Matei, L., Buliga, V., Groza, M., Burger, A., & Mandal, K. C. (2013). $Cd_{0.9}Zn_{0.1}Te$ crystal growth and fabrication of large volume single-polarity charge sensing gamma detectors. *IEEE Transactions on Nuclear Science, 60*, 2853–2858.
25. Krishna, R. M., Chaudhuri, S. K., Zavalla, K. J., & Mandal, K. C. (2013). Characterization of $Cd_{0.9}Zn_{0.1}Te$ based virtual Frisch grid detectors for high energy gamma ray detection. *Nuclear Instruments and Methods in Physics Research A, 701*, 208–213.
26. Chaudhuri, S. K., Zavalla, K. J., Krishna, R. M., & Mandal, K. C. (2013). Biparametric analyses of charge trapping in $Cd_{0.9}Zn_{0.1}Te$ based virtual Frisch grid detectors. *Journal of Applied Physics, 113*, 074504.
27. Chaudhuri, S. K., Kleppinger, J. W., Karadavut, O. F., Nag, R., & Mandal, K. C. (2021). Quaternary semiconductor $Cd_{1-x}Zn_xTe_{1-y}Se_y$ for high-resolution, room-temperature gamma-ray detection. *Crystals, 11*, 827.
28. Chaudhuri, S. K., Sajjad, M., Kleppinger, J. W., & Mandal, K. C. (2020). Charge transport properties in CdZnTeSe semiconductor room-temperature γ-ray detectors. *Journal of Applied Physics, 127*, 245706.
29. Yakimov, A., Smith, D., Choi, J., & Araujo, S. (2019). Growth and characterization of detector-grade CdZnTeSe by horizontal Bridgman technique. *Proceedings of SPIE, 1114*, 11141N.
30. Egarievwe, S. U., Roy, U. N., Agbalagba, E. O., Harrison, B. A., Goree, C. A., Savage, E. K., & James, R. B. (2020). Optimizing CdZnTeSe Frisch-grid nuclear detector for gamma-ray spectroscopy. *IEEE Access, 8*, 137530–137539.
31. Pipek, J., Betušiak, M., Belas, E., Grill, R., Praus, P., Musiienko, A., Pekarek, J., Roy, U. N., & James, R. B. (2021). Charge transport and space-charge formation in $Cd_{1-x}Zn_xTe_{1-y}Se_y$ radiation detectors. *Physical Review Applied, 15*, 054058.
32. Roy, U. N., Camarda, G. S., Cui, Y., & James, R. B. (2021). Optimization of selenium in CdZnTeSe quaternary compound for radiation detector applications. *Applied Physics Letters, 118*, 152101.
33. Hwang, S., Yu, H., Bolotnikov, A. E., James, R. B., & Kim, K. (2019). Anomalous Te inclusion size and distribution in CdZnTeSe. *IEEE Transactions on Nuclear Science, 66*, 2329–2332.
34. Roy, U. N., Camarda, G. S., Cui, Y., Gul, R., Hossain, A., Yang, G., Zazvorka, J., Dedic, V., Franc, J., & James, R. B. (2019). Role of selenium addition to CdZnTe matrix for room-temperature radiation detector applications. *Scientific Reports, 9*, 1620.
35. Roy, U. N., Camarda, G., Cui, Y., Yang, G., & James, R. B. (2021). Impact of selenium addition to the cadmium zinc telluride matrix for producing high energy resolution X and gamma ray detectors. *Scientific Reports, 11*, 10338.
36. Kleppinger, J. W., Chaudhuri, S. K., Roy, U. N., James, R. B., & Mandal, K. C. (2021). Growth of $Cd_{0.9}Zn_{0.1}Te_{1-y}Se_y$ single crystals for room temperature gamma-ray detection. *IEEE Transactions on Nuclear Science, 68*, 2429–2434.
37. Gul, R., Roy, U. N., Camarda, G. S., Hossain, A., Yang, G., Vanier, P., Lordi, V., Varley, J., & James, R. B. (2017). A comparison of point defects in $Cd_{1-x}Zn_xTe_{1-y}Se_y$ crystals grown by Bridgman and traveling heater methods. *Journal of Applied Physics, 121*, 125705.
38. Chaudhuri, S. K., Sajjad, M., & Mandal, K. C. (2020). Pulse-shape analysis in $Cd_{0.9}Zn_{0.1}Te_{0.98}Se_{0.02}$ room-temperature radiation detectors. *Applied Physics Letters, 116*, 162107.
39. Rejhon, M., Franc, J., Dědič, V., Pekárek, J., Roy, U. N., Grill, R., & James, R. B. (2018). Influence of deep levels on the electrical transport properties of CdZnTeSe detectors. *Journal of Applied Physics, 124*, 235702.
40. Chaudhuri, S. K., Sajjad, M., Kleppinger, J. W., & Mandal, K. C. (2020). Correlation of space charge limited current and γ-ray response of $Cd_xZn_{1-x}Te_{1-y}Se_y$ room-temperature radiation detectors. *IEEE Electron Device Letters, 41*, 1336–1339.

41. Rejhon, M., Dědič, V., Beran, L., Roy, U. N., Franc, J., & James, R. B. (2019). Investigation of deep levels in CdZnTeSe crystal and their effect on the internal electric field of CdZnTeSe gamma-ray detector. *IEEE Transactions on Nuclear Science, 66*, 1952–1958.
42. Chaudhuri, S. K., Kleppinger, J. W., Karadavut, O., Nag, R., Panta, R., Agostinelli, F., Sheth, A., Roy, U. N., James, R. B., & Mandal, K. C. (2022). Synthesis of CdZnTeSe single crystals for room temperature radiation detector fabrication: Mitigation of hole trapping effects using a convolutional neural network. *Journal of Materials Science: Materials in Electronics, 33*, 1452–1463.
43. Soundararajan, R., & Lynn, K. G. (2012). Effects of excess tellurium and growth parameters on the band gap defect levels in $Cd_xZn_{1-x}Te$. *Journal of Applied Physics, 112*, 073111.
44. Yang, G., Jie, W., Li, Q., Wang, T., Li, G., & Hua, H. (2005). Effects of In doping on the properties of CdZnTe single crystals. *Journal of Crystal Growth, 283*, 431–437.
45. Harrison, M. J., Graebner, A. P., McNeil, W. J., & McGregor, D. S. (2006). Carbon coating of fused silica ampoules. *Journal of Crystal Growth, 290*, 597–601.
46. Nag, R., Chaudhuri, S. K., Kleppinger, J. W., Karadavut, O., & Mandal, K. C. (2021). Characterization of vertical Bridgman grown $Cd_{0.9}Zn_{0.1}Te_{0.97}Se_{0.03}$ single crystal for room-temperature radiation detection. *Journal of Materials Science: Materials in Electronics, 32*, 26740–26749.
47. Nag, R., Chaudhuri, S. K., Kleppinger, J. W., Karadavut, O., & Mandal, K. C. (2022). Vertical gradient freeze growth of detector grade CdZnTeSe single crystals. *Journal of Crystal Growth, 596*, 126826.
48. Bolotnikov, A. E., Abdul-Jaber, N. M., Babalola, O. S., Camarda, G. S., Cui, Y., Hossain, A. M., Jackson, E. M., Jackson, H. C., James, J. A., Kohman, K. T., Luryi, A. L., & James, R. B. (2008). Effects of Te inclusions on the performance of CdZnTe radiation detectors. *IEEE Transactions on Nuclear Science, 55*, 2757–2764.
49. Brovko, A., Amzallag, O., Adelberg, A., Chernyak, L., Raja, P. V., & Ruzin, A. (2021). Effects of oxygen plasma treatment on $Cd_{1-x}Zn_xTe$ material and devices. *Nuclear Instruments and Methods in Physics Research B, 1004*, 165343.
50. Sajjad, M., Chaudhuri, S. K., Kleppinger, J. W., & Mandal, K. C. (2020). Growth of large-area $Cd_{0.9}Zn_{0.1}Te$ single crystals and fabrication of pixelated guard-ring detector for room-temperature γ-ray detection. *IEEE Transactions on Nuclear Science, 67*, 1946–1951.
51. Hecht, K. (1932). Zum mechanismus des lichtelektrischen primärstromes in isolierenden kristallen. *Zeitschrift für Physik, 77*, 235–245.
52. Sellin, P. J., Davies, A. W., Lohstroh, A., Ozsan, M. E., & Parkin, J. (2005). Drift mobility and mobility-lifetime products in CdTe:Cl grown by the travelling heater method. *IEEE Transactions on Nuclear Science, 52*, 3074–3078.
53. Zhang, S. B., Wei, S. H., Zunger, A., & Katayama-Yoshida, H. (1998). Defect physics of the $CuInSe_2$ chalcopyrite semiconductor. *Physical Review B, 57*, 9642–9656.
54. Freysoldt, C., Neugebauer, J., & Van de Walle, C. G. (2009). Fully *ab initio* finite-size corrections for charged-defect supercell calculations. *Physical Review Letters, 102*, 016402.
55. Perdew, J. P., Ernzerhof, M., & Burke, K. (1996). Rationale for mixing exact exchange with density functional approximations. *The Journal of Chemical Physics, 105*, 9982–9985.
56. Lang, D. V. (1974). Deep-level transient spectroscopy: A new method to characterize traps in semiconductors. *Journal of Applied Physics, 45*, 3023–3032.
57. Tapiero, M., Benjelloun, N., Zielinger, J. P., El Hamd, S., & Noguet, C. (1988). Photoinduced current transient spectroscopy in high-resistivity bulk materials: Instrumentation and methodology. *Journal of Applied Physics, 64*, 4006–4012.
58. Hurtes, C., Boulou, M., Mitonneau, A., & Bois, D. (2008). Deep-level spectroscopy in high-resistivity materials. *Applied Physics Letters, 32*, 821.
59. Mandal, K. C., Muzykov, P. G., Chaudhuri, & Terry, J. R. (2013). Low energy X-ray and γ-ray detectors fabricated on n-type 4H-SiC epitaxial layer. *IEEE Transactions on Nuclear Science, 60*, 2888–2893.
60. Mandal, K. C., Kleppinger, J. W., & Chaudhuri, S. K. (2020). Advances in high-resolution radiation detection using 4H-SiC epitaxial layer devices. *Micromachines, 11*(3), 254.

Charge Sharing and Cross Talk Effects in High-Z and Wide-Bandgap Compound Semiconductor Pixel Detectors

Antonino Buttacavoli, Fabio Principato, Gaetano Gerardi, Manuele Bettelli, Matthew C. Veale, and Leonardo Abbene

1 Introduction

The development of photon-counting detectors with energy resolving capabilities opened exciting perspectives in the field of X-ray imaging, with strong impacts in several applications, from diagnostic/nuclear medicine, synchrotron science to non-destructive testing (NDT) in food industry [1–9]. Energy-resolved photon-counting (ERPC) detectors are typically based on direct semiconductor materials, with particular emphasis on high-Z and wide-bandgap compound semiconductors [10–14] for enhancements in detection efficiency and energy resolution near room-temperature conditions. Currently, cadmium telluride (CdTe) and cadmium–zinc–telluride (CdZnTe or CZT) give the best performance [1–15]. At low X-ray energies (<100 keV), unsurpassed energy resolution (<1.5% FWHM at 60 keV) and high detection efficiency (>90%) are obtained with thin CdTe pixel detectors (thickness <1 mm) fabricated with Schottky electrical contacts [6, 12–15]. However, the increase of the bias-induced polarization effects with the detector thickness [6, 16–19] limited the use of thick Schottky CdTe detectors for high energies. At energies greater than 100 keV, interesting energy resolution (<2% FWHM at 122 keV) [4, 20, 21] is obtained with thick CZT pixel detectors with quasi-ohmic electrical contacts that are immune to the bias-induced polarization effects. CdTe/CZT detectors with

A. Buttacavoli · F. Principato · G. Gerardi · L. Abbene (✉)
Department of Physics and Chemistry (DiFC)-Emilio Segrè, University of Palermo, Palermo, Italy
e-mail: leonardo.abbene@unipa.it

M. Bettelli
IMEM/CNR Parma, Parma, Italy

M. C. Veale
Science and Technology Facilities Council, Rutherford Appleton Laboratory, Chilton, UK

© The Author(s), under exclusive license to Springer Nature Switzerland AG 2023
L. Abbene, K. (Kris) Iniewski (eds.), *High-Z Materials for X-ray Detection*,
https://doi.org/10.1007/978-3-031-20955-0_10

sub-millimetre pixels represent a key choice for both spatial and energy resolution improvements, in agreement with the *small pixel effect* [10]. The pixelated anodes are characterized by electron-sensing properties (*small pixel effect*), which are very important to minimize the spectral distortions related to the poor hole transport properties of CdTe/CZT materials. However, when small pixel arrays are used, the presence of multiple coincidence events among neighbouring pixels, due to charge sharing and cross talk phenomena [22–24], represents a serious critical issue. The goal of this chapter is to present an overview of the effects of these phenomena on the spectroscopic response of CZT pixel detectors, with particular attention to the current state-of-art of the main discrimination and correction techniques. Moreover, the results from original energy-recovery procedures of multiple charge sharing events, recently developed by our group, are also shown.

2   Charge Sharing and Cross Talk Events in CZT Pixel Detectors

Nowadays, CZT detectors have reached superb maturity level in room temperature X-ray and gamma-ray detection, from energies of few keV up to 1 MeV [1–11, 25–28]. After the first CZT detector was successfully developed by Butler in 1992 [29], intense research activities started with important progresses in both crystal growth and electrical contact technology. Typically, the best spectroscopic grade CZT crystals are grown through the Bridgman (B) [5, 30–32] and the traveling heater method (THM) growth [33–36] techniques. In particular, recent progresses in the charge carrier transport properties in THM-CZT crystals have been made. The development of THM-CZT detectors with high electron transport properties, i.e. characterized by mobility-lifetime products of electrons $\mu_e \tau_e$ greater than 10^{-2} cm^2/V, was pioneered by Chen in 2007 [36]. Since then, several suppliers (Redlen Technologies, Canada; Kromek, UK) were able to fabricate high $\mu_e \tau_e$ CZT crystals with thicknesses greater than 10 mm. Great efforts have been also made to enhance the mobility-lifetime products $\mu_h \tau_h$ of the holes, especially for high-flux measurements. Improvements in hole charge transport properties are necessary to minimize the effects of radiation-induced polarization phenomena at high fluxes [37–40]. Recently, high $\mu_h \tau_h$ THM-CZT crystals ($\mu_h \tau_h > 10^{-4}$ cm^2/V) are fabricated by Redlen for high-flux applications [41–44]. High $\mu_h \tau_h$ CZT detectors (high-flux HF-CZT detectors) are typically used for high-flux measurements, while high $\mu_e \tau_e$ CZT materials (low-flux LF-CZT) for thick electron-sensing detectors generally working at low-flux conditions. As well-known, charge sharing and cross talk represent the main critical issues in sub-millimetre CZT pixel detectors, and intense investigations have been made on the mitigation of the effects on the performance of the detectors [22–24, 45–49]. Typically, these phenomena mainly occur in the region near the inter-pixel gaps of the detector arrays, and, with a dominant photoelectric interaction, they are due to charge sharing and fluorescence/weighting potential cross talk events. In the following sections, we will present the typical events contributing on the coincidences among pixels and their effects on CZT pixel/strip detectors. We investigated on LF/HF THM CZT pixel detectors with the same geometrical layout (3 × 3 pixel arrays with pixel pitches of 500/250 μm) and the same digital pulse processing electronics [50, 51].

2.1 Charge Sharing Events

Charge sharing is simply referred to the splitting of the charge cloud created by a single interacting event and collected by two or more pixels. The pulses from charge sharing are *collected-charge pulses*, i.e. generated by a real collection of the charge by the pixels. In CZT pixel detectors, the charge cloud collected by the pixels is mainly due to the drift of the electrons, in agreement with the *small pixel effect* [10]. The region covered by the broadened electron cloud depends upon charge diffusion, Coulomb charge repulsion, K-shell X-ray fluorescence and Compton scattering. Figure 1a shows the results from a simulation of the evolution of the electron charge cloud size (FWHM) in a CZT detector vs. the drift distance, generated by photoelectrons in the energy range of 25–662 keV.

The simulation involves the physical processes of Coulomb repulsion and charge diffusion [52]. The results are presented at (a) 5000 V/cm electric field, representing the current state-of-art of electric field in CZT detectors, while (b) the electric field of 10,000 V/cm was recently obtained in new high-bias voltage CZT pixel detectors [30, 37]. For example, at 60 keV the size of the charge cloud is about 60 μm and 45 μm over a drift distance of 2 mm, with 5000 V/cm and 10,000 V/cm, respectively. The pulses from charge-sensitive preamplifiers, related to a typical charge sharing event (60 keV) involving four pixels (event multiplicity $m = 4$), are shown in Fig. 1c. The broadening of the charge cloud size can be also increased by the presence of fluorescent X-rays, which emission probability is very high in CZT materials (~85% of all photoelectric absorptions) [53, 54]. At energies greater than the K-shell absorption energy of the CZT material (26.7 keV, 9.7 keV and 31.8 keV for Cd, Zn and Te, respectively), the emission of fluorescent X-rays of 23.2 keV (Cd-$K_{\alpha 1}$) and 27.5 keV (Te-$K_{\alpha 1}$) is more probable, with attenuation lengths of 116 μm and 69 μm, respectively [55]. As shown in our previous works [20, 55], 2-mm-thick detectors with pixel pitches of 250 μm and inter-pixel gap of 50 μm showed an increase of the coincidence percentage from 50% to 80% at 22 keV and 60 keV, respectively; this demonstrates as the fluorescent X-rays, present at 60 keV, give a great contribute on charge sharing.

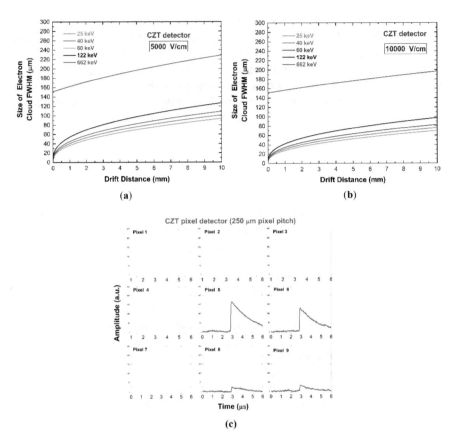

Fig. 1 Simulation of the evolution of the electron cloud size (FWHM) vs. the drift distance in CZT detectors generated by photoelectrons in the energy range of 25–662 keV; Coulomb repulsion and charge diffusion are used in the simulation [52]. (**a**) Electric field of 5000 V/cm and (**b**) 10,000 V/cm. (**c**) The typical detector/preamplifier pulses from charge sharing of a single 59.5 keV photon interaction involving four pixels. The pulses are in temporal coincidence within a coincidence time window (CTW) of 10 ns

2.2 Fluorescence Cross Talk Events

Fluorescent X-rays, escaping from a pixel, can also produce cross talk events on neighbouring pixels. These cross talk events are collected-charge pulses, i.e. created by the charge carriers really collected by the pixels. Figure 2 shows an example of fluorescent cross talk events between two adjacent pixels of a CZT pixel detector. In particular, we measured the double temporal coincidences ($m = 2$) of two adjacent pixels for a collimated irradiation ($10 \times 10 \ \mu m^2$) with synchrotron X-rays at the centre of one of the two pixels (pixel 5 of Fig. 2). These measurements were performed at the B16 test beamline of the Diamond Light Source synchrotron (Didcot, UK; http://www.diamond.ac.uk/Beamlines/Materials/B16).

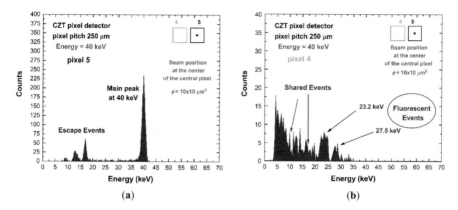

Fig. 2 Example of fluorescence cross talk events between two adjacent pixels of a HF-THM CZT pixel detector. A collimated X-ray synchrotron beam (10 × 10 μm^2) was irradiated at the centre of the pixel no. 5, and the energy spectra at 40 keV of the two adjacent pixels were measured: (**a**) pixel 5, (**b**) pixel 4. The presence of fluorescence cross talk events is clearly visible

Here, the energy spectra of the central pixel (Fig. 2a) and the adjacent pixel (Fig. 2b) are shown. All events of the adjacent pixel (pixel no. 4) are in double temporal coincidence with events of the central pixel (pixel no. 5); the fluorescent events are clearly visible.

2.3   Weighting Potential Cross Talk Events

Cross talk events can be also created by *induced-charge pulses* generated on neighbouring non-collecting pixels [20, 24, 26, 48]. These pulses, also called *transient pulses*, are due to the weighting potential cross talk. To explain the physical origin and the shape of these pulses, a brief description of the detector signal formation is necessary. Generally, the generation of pulses in CZT detectors can be clearly explained through the Shockley–Ramo theorem [10], with the concept of weighting potential. The charge generated on a pixel is related to the variation of the weighting potential between the charge generation and the collection points. The presence of non-monotonic weighting potentials on non-collecting pixels creates induced-charge pulses, generated by charge carriers really collected by another pixel (the collecting pixel). Therefore, the movement of charge carriers in the vicinity of a collecting pixel can induce a small charge on the surrounding non-collecting pixel. This effect is known as weighting potential cross talk. Ideally, the induced-charge pulses are fast bipolar pulses that in the absence of charge trapping will drop quickly to zero amplitude. However, the presence of charge trapping and the different interaction depths of the events can create transient pulses with positive or negative polarities [20, 26]. Figure 3a shows a typical positive collected-charge

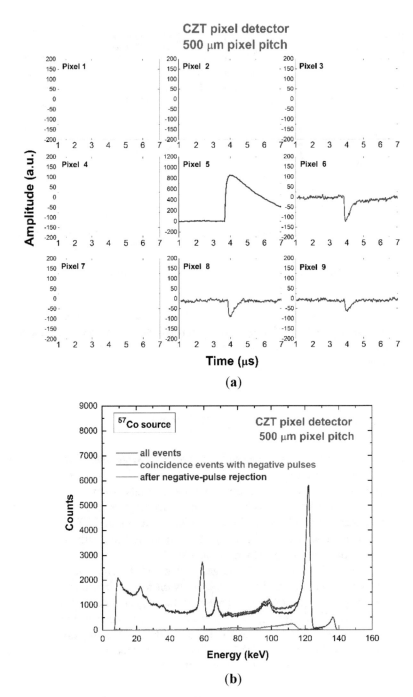

Fig. 3 (**a**) Example of negative induced-charge pulses (pixels no. 6, 8 and 9) measured in temporal coincidence with a positive collected-charge pulse (pixel 5). (**b**) The raw spectrum of the central pixel 5 (blue line), the spectrum of the coincidence events with the negative pulses of the eight adjacent pixels (pink line) and after their rejection (brown line). This result clearly shows as the negative pulses are created by photon interactions near the pixel anode, where the positive collected-charge pulses suffer from incomplete charge collection due to the hole trapping

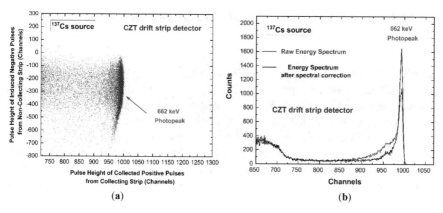

Fig. 4 Example of correction of incomplete charge collection by using the negative-induced charge pulses. (**a**) Two-dimensional (2D) scatter plot of the negative heights of the pulses from a non-collecting drift versus the height of pulses from an anode collecting strip. The curvature highlights the presence of pulses with incomplete charge collection related to high values of the negative heights. (**b**) Measured ^{137}Cs energy spectra after the spectral correction (black line) [26, 27]. The events in the tailing of the photopeak, i.e. characterized by incomplete charge collection, are well detected and corrected

pulse (brown line) from the collecting central pixel of a 3×3 CZT pixel array. This pulse is in temporal coincidence with negative induced-charge pulses (blue lines) of adjacent pixels. The negative pulses are induced-charge pulses produced by non-collecting pixels, and they are mainly related to photon interactions near the pixel boundary and at interaction depths close to the pixel plane (i.e. the pixelated anode). For interactions just outside the pixel boundary and at depths near the pixelated anode, the collecting pixel will give a positive collected-charge pulse, which will be in temporal coincidence with a negative induced-charge pulse generated by the adjacent pixels. The monotonic weighting potential of the collecting pixel will give a positive pulse, even if characterized by charge losses, while the adjacent pixels (i.e. the non-collecting pixels), due to the non-monotonic behaviour of the weighting potential and to the hole trapping, give fast transient pulses with negative polarity. Therefore, these negative pulses are in temporal coincidence with positive pulses characterized by charge losses due to hole trapping, as clearly shown in Fig. 3b.

Moreover, as demonstrated in Fig. 4, it is possible to correct incomplete charge collection by modelling the behaviour of the positive/negative heights [26, 27]. In this case, we presented an example on a LF-THM CZT strip detector, characterized by electron-sensing weighting potential, as the pixel detector one.

3   Detection and Correction of Charge Sharing in CZT Pixel Detectors

In this section, we will present the effects of charge sharing and fluorescence cross talk events in the energy spectra of CZT pixel detectors (3×3 CZT arrays with pixel pitches of 500/250 μm). Particular attention will be given to the presentation of interesting correction techniques of multiple charge sharing events (multiplicity $m \geq 2$). We will start with the presentation of the energy spectra typically showed after charge sharing investigations. In particular, Fig. 5 shows three different energy spectra of uncollimated 59.5 keV photons from ^{241}Am source, related to the central pixel of the 250 μm array of a 2-mm-thick CZT detector. The brown line represents the energy spectrum of all events (raw spectrum), and the blue line is the spectrum of the events in temporal coincidence with all eight neighbouring pixels.

The percentage of coincidence events is very high (80%), and the typical distortions related to charge sharing and fluorescent cross talk are clearly visible: the fluorescent peaks at 23.2 keV and 27.5 keV, the escape peaks at 36.3 keV and 32 keV, the low-energy background and tailing. The black line is the spectrum after charge sharing discrimination (CSD), i.e. after the rejection of coincidence events

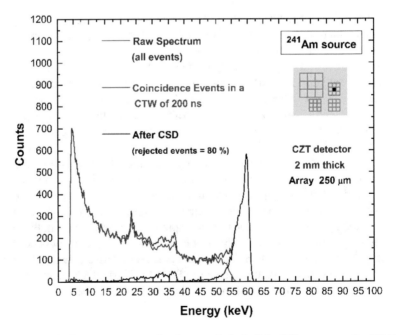

Fig. 5 Charge sharing measurements for the central pixel of the 250 μm array of a HF-THM CZT pixel detector. The black line represents the uncollimated ^{241}Am spectrum after charge sharing discrimination (CSD). The raw spectrum (brown line) of all events and the spectrum of the coincidence events with all eight neighbouring pixels (blue line) are also shown. The yellow inset shows the layout of the anode of the CZT pixel detector

(80%). The CSD works well allowing the rejection of all charge sharing distortions, even if it produces a strong reduction of the pixel throughput and counting efficiency (80% rejected events). The tailing below the main photopeak is not mitigated; these events are not in temporal coincidence with neighbouring pixels, due to their energies below the detection energy threshold (4 keV). Moreover, escape peaks are also present in the spectra even after CSD, due to fluorescence events escaping from the cathode or absorbed on the guard-ring. CSD represents the state of art on the mitigation of charge sharing effects; however, the high number of rejected events, often greater than 50% of all detected events, gives a strong reduction of the throughput and the counting efficiency of the detectors. To avoid this, the rejected events after CSD can be recovered through the charge sharing addition (CSA) technique, which consists of summing the energies of the coincidence events. Unfortunately, as documented in the literature [20, 45, 49, 55], the energy recovered after CSA (E_{CSA}) is often lower than true photon energy, due to the presence of charge losses near the region of inter-pixel gap. These losses are related to the presence of distorted electric field lines at the inter-pixel gap [22]; in fact, due to the surface conductivity at the inter-pixel gap, the electric field lines can intersect the surface of the inter-pixel gaps, where some charges can be trapped. Generally, the charge losses depend on the interaction position within the gap, with high effects near the centre. Now, we will present some techniques able to correct charge losses after CSA even when multiple pixels are involved.

3.1 Coincidence Events with Multiplicity m = 2

The predominant contribution to the overall coincidence events is represented by the events with multiplicity $m = 2$. Double coincidence events are mainly due to charge sharing events near the inter-pixel gap, fluorescent cross talk, and mixed shared/fluorescent events. Figure 6 shows the results obtained after the application of CSA in CZT pixel detectors, at energies below (^{109}Cd source) and above (^{241}Am source) the K-shell absorption energy of CZT material. Energy deficits in the spectra after CSA are observed.

Moreover, the distribution of these energy deficits within the inter-pixel gap between two pixels is also presented in Fig. 7. The two-dimensional (2D) scatter plots show the summed energy E_{CSA} of the coincidence events ($m = 2$) between two adjacent pixels (pixels no. 5 and no. 6), after CSA, versus the charge sharing ratio R. The quantity R, calculated from the ratio between the energies of two adjacent pixels, gives indications about the position of the photon interaction between the two pixels. The curvature shows the presence of energy deficits at all energies; they are more severe at $R = 0$, i.e. related to events interacting at the centre of the inter-pixel gap. The summed energy E_{CSA} of some coincidence events at about $R = 0.22$ is fully recovered after CSA. These events represent the couples of the fluorescent X-rays at 23.2 keV and the escape events at 36.3 keV. To recover the energy of double coincidences, two separate procedures for adjacent and diagonal pixels can be used.

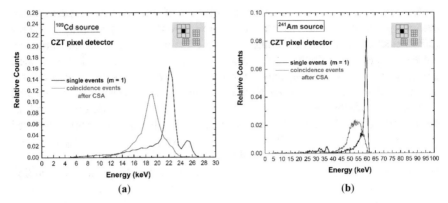

Fig. 6 Example of the presence of charge losses after charge sharing addition (CSA) in LF-THM CZT pixel detector. The measured (**a**) [109]Cd and (**b**) [241]Am energy spectra after the application of CSA. The energy spectra of the single events (black lines) and the spectra of the coincidence events with the eight adjacent pixels (multiplicity $m = 2$) after CSA (red lines). The energy spectra after CSA are characterized by energy deficits due to the presence of charge losses near the inter-pixel gaps

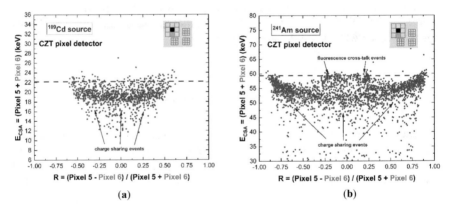

Fig. 7 Example of the presence of charge losses after charge sharing addition (CSA). Two-dimensional (2D) scatter plot of the summed energy E_{CSA} of the coincidence events ($m = 2$) between two adjacent pixels (the central pixel no. 5 and the pixel no. 6), after CSA, versus the charge sharing ratio R. The plots, (**a**) at 22.1 keV and (**b**) at 59.5 keV, highlight the dependence of the energy deficit on the position within the inter-pixel gap

The double coincidence events between adjacent pixels are mainly dominated by charge shared events and by a small contribute of fluorescent cross talk events. The energy of double shared events can be recovered through the correction technique, termed double charge sharing correction (double CSC), presented in our previous work [20].

The technique exploits the strong relation between the energy E_{CSA} after CSA and the charge sharing ratio R (Fig. 7). The relation between the energy E_{CSA} and R was modelled through the following equation:

$$E_{CSA} = E - \Delta E_{CSA}(0) \cdot \left(1 - R^2\right), \tag{1}$$

where E is the true photon energy and $\Delta E_{CSA}(0) = E - E_{CSA}(0)$ is the energy lost at the centre of the inter-pixel gap, i.e. at $R = 0$. Moreover, we demonstrated [20] a linear behaviour of the energy lost $\Delta E_{CSA}(0)$ with the true energy E:

$$\Delta E_{CSA}(0) = k_1 + k_2 \cdot E, \tag{2}$$

where k_1 and k_2 are the slope and the y-intercept of the linear function. Combining Eqs. (1) and (2) yields:

$$E_{corr} = E = \left[\frac{E_{CSA}(R) + k_1 \cdot \left(1 - R^2\right)}{1 - k_2 \cdot \left(1 - R^2\right)} \right], \tag{3}$$

By using Eq. (3), it is possible to correct the charge losses after CSA through the measurement of the bi-parametric distribution $E_{CSA} - R$ and the estimation of the constants k_1 and k_2, which can be obtained by a preliminary calibration procedure. We stress that this correction does not depend on the photon energy, but it is related to the physical and geometrical characteristics of the detectors. Figure 8 presents the results after the application of standard CSA and double CSC. After double CSC, the energy loss is recovered, and the energy resolution is also improved.

At energies greater than the K-shell absorption energy of the CZT material, the double coincidence events between adjacent pixels also contain fluorescent cross talk events, which can be easily corrected after standard CSA. The selection of these events (fluorescent event and escape peak event) is simple for mono-energetic X-ray sources, but it is challenge for poly-energetic sources. Concerning the double coincidence events between diagonal pixels, we observed that they are due to pure fluorescence cross talk events. This is clearly shown in Fig. 9. In particular, the 2D scatter plot of Fig. 9a highlights the energy recovery of the energy after standard CSA at about $R = \pm 0.22$ and $R = \pm 0.076$, which are related to the fluorescent X-rays of 23.2 and 27.5 keV, respectively ([241]Am source). The same result is also confirmed through the energy spectrum after CSA (Fig. 9b). At energies below the K-shell absorption energy of CZT (e.g. by using[109]Cd source), no double coincidence events were observed between diagonal pixels.

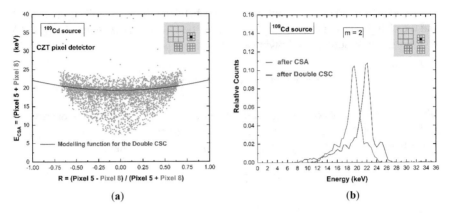

Fig. 8 Correction of the charge losses after CSA for double coincidence events ($m = 2$). (**a**) 2D scatter plot of the summed energy of the coincidence events ($m = 2$) between the pixel no. 5 and pixel no. 8, after CSA, versus the ratio R (uncollimated [109]Cd source). The blue line is the modelling function (Eq. 1) [20], used to correct charge losses. (**b**) The energy spectra after CSA (red line) and after double CSC (blue line). The complete recovery of the energy after double CSC and improvements of the energy resolution are clearly visible

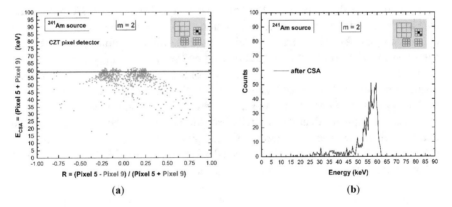

Fig. 9 Correction of the fluorescence cross talk events between diagonal pixels. (**a**) 2D scatter plot of the summed energy of the coincidence events ($m = 2$) between diagonal pixels (no. 5 and pixel no. 9), after CSA, versus the ratio R. (**b**) The energy spectrum shows the absence of charge losses after CSA, demonstrating as the double coincidence events between diagonal pixels are mainly due to the fluorescent/escape events

4   Coincidence Events with Multiplicity $m > 2$

The presence of coincidence events with multiplicity $m > 2$ is also observed in CZT pixel detectors. These events are created by mixed fluorescent/sharing coincidence events at the inter-pixel gap. In particular, some triple coincidence events (i.e. $m = 3$) can be often obtained from a true quadruple coincidence, where the energy of the pulse of one pixel is below the detection energy threshold, e.g. of 4 keV in our case.

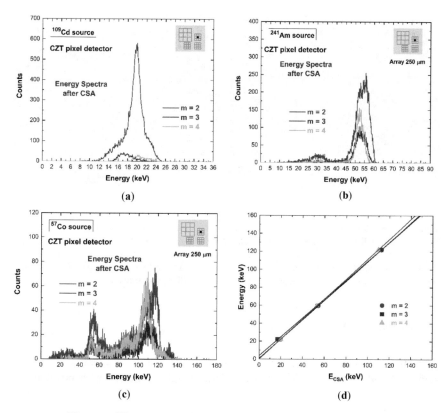

Fig. 10 (**a**) ^{109}Cd, (**b**) ^{241}Am and (**c**) ^{57}Co spectra after CSA at various multiplicities. (**d**) The linearity of E_{CSA} with the true photon energy opens up to a correction of charge losses after CSA, through a simple energy re-calibration procedure

The recovery of the energy of multiple coincidence events is still challenging. In Fig. 10, we present the energy spectra of coincidence events after CSA at different multiplicities and energies. All energy spectra suffer from charge losses after CSA. However, the linear behaviour of the summed energy E_{CSA} with the true photon energy (Fig. 10d) allows the recovery of the energy through a simple energy re-calibration procedure. We point out as the 4 keV intercept of the linear function for $m = 3$ (the brown line of Fig. 10d) highlights the possible measurement of triple coincidence events from true quadruple coincidences, where the energy of one of the four pixels is below the detection threshold (4 keV) and, therefore, not detected.

Finally, we applied the various proposed correction techniques for each related multiplicity m. In particular, we used the double CSC technique on $m = 2$ events between adjacent pixels, the standard CSA for $m = 2$ events between diagonal pixels and the re-calibrated CSA for both $m = 3$ and $m = 4$ events. The complete correction is termed multiple CSC technique. The results are shown in Fig. 11.

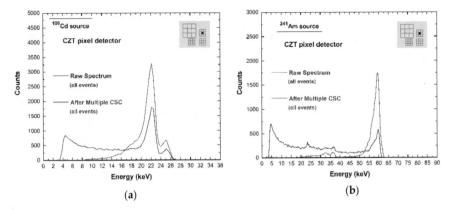

Fig. 11 Examples of charge loss correction of multiple coincidence events. (**a, b**) The raw energy spectra (brown lines) and the corrected spectra (blue lines) after the multiple charge sharing correction (CSC). We applied the double CSC technique on $m = 2$ events between adjacent pixels, the standard CSA for $m = 2$ events between diagonal pixels and the re-calibrated CSA for both $m = 3$ and $m = 4$ events. The energy resolution was also slightly improved

5 Conclusions

The effects of charge sharing and cross talk events in CZT pixel detectors are reviewed and discussed. Collected and induced-charge pulses with positive and negative polarities were observed. The correction of incomplete charge collection through the negative induced-charge pulses is presented and successfully applied in CZT pixel/strip detectors. Original techniques able to correct the charge losses after charge sharing addition (CSA) are also presented. Different approaches were used for adjacent and diagonal pixels, taking into account the number of involved pixels (i.e. the multiplicity m). One approach, exploiting the relation between the summed energy after CSA and the charge sharing ratio R of double coincidence events ($m = 2$), allowed the recovery of charge losses between adjacent pixels. A second technique, based on the linear behaviour of charge losses after CSA with the true photon energy, was also implemented to reconstruct multiple coincidence events with $m > 2$. The energy of double coincidence events between diagonal pixels, mainly due to fluorescence cross talk events, was successfully recovered after the standard CSA. Results on CZT pixel detectors showed improved counting efficiency compared to using only isolated events (i.e. after CSD) and improved energy resolution compared to the standard CSA techniques.

References

1. Barber, W. C., Wessel, J. C., Nygard, E., & Iwanczyk, J. S. (2015). Energy dispersive CdTe and CdZnTe detectors for spectral clinical CT and NDT applications. *Nuclear Instruments and Methods A, 784*, 531–537.
2. Del Sordo, S., Strazzeri, M., Agnetta, G., Biondo, B., Celi, F., Giarrusso, S., Mangano, A., Russo, F., Caroli, E., Donati, A., et al. (2004). Spectroscopic performances of 16 × 16 pixel CZT imaging hard-X-ray detectors. *Nuovo Cimento B, 119*, 257–270.
3. Iwanczyk, J., Nygard, E., Meirav, O., Arenson, J., Barber, W. C., Hartsough, N. E., Malakhov, N., & Wessel, J. C. (2009). Photon counting energy dispersive detector arrays for X-ray imaging. *IEEE Transactions on Nuclear Science, 56*, 535–542.
4. Seller, P., Bell, S., Cernik, R. J., Christodoulou, C., Egan, C. K., Gaskin, J. A., Jacques, S., Pani, S., Ramsey, B. D., Reid, C., et al. (2011). Pixellated Cd(Zn)Te high-energy X-ray instrument. *Journal of Instrumentation, 6*, C12009.
5. Szeles, C., Soldner, S. A., Vydrin, S., Graves, J., & Bale, D. S. (2008). CdZnTe semiconductor detectors for spectroscopic X-ray imaging. *IEEE Transactions on Nuclear Science, 55*, 572–582.
6. Abbene, L., Gerardi, G., & Principato, F. (2015). Digital performance improvements of a CdTe pixel detector for high flux energy-resolved X-ray imaging. *Nuclear Instruments and Methods A, 777*, 54–62.
7. Iniewski, K. (2014). CZT detector technology for medical imaging. *Journal of Instrumentation, 9*, C11001.
8. Nguyen, J., Rodesch, P. A., Richtsmeier, D., Iniewski, K., & Bazalova-Carter, M. (2021). Optimization of a CZT photon counting detector for contaminant detection. *Journal of Instrumentation, 16*, P11015.
9. Richtsmeier, D., Guliyev, E., Iniewski, K., & Bazalova-Carter, M. (2021). Contaminant detection in non-destructive testing using a CZT photon-counting detector. *Journal of Instrumentation, 16*, P01011.
10. Del Sordo, S., Abbene, L., Caroli, E., Mancini, A. M., Zappettini, A., & Ubertini, P. (2009). Progress in the development of CdTe and CdZnTe semiconductor radiation detectors for astrophysical and medical applications. *Sensors, 9*, 3491–3526.
11. Owens, A. (2006). Semiconductor materials and radiation detection. *Journal of Synchrotron Radiation, 13*, 143–150.
12. Sammartini, M., Gandola, M., Mele, F., Garavelli, B., Macera, D., Pozzi, P., & Bertuccio, G. (2021). X–γ -ray spectroscopy with a CdTe pixel detector and SIRIO preamplifier at deep sub microsecond signal-processing time. *IEEE Transactions on Nuclear Science, 68*, 70–75.
13. Sammartini, M., Gandola, M., Mele, F., Garavelli, B., Macera, D., Pozzi, P., & Bertuccio, G. (2018). A CdTe pixel detector–CMOS preamplifier for room temperature high sensitivity and energy resolution X and γ ray spectroscopic imaging. *Nuclear Instruments and Methods A, 910*, 168–173.
14. Wilson, M. D., Bell, S. J., Cernik, R. J., Christodoulou, C., Egan, C. K., O'Flynn, D., Jacques, S., et al. (2013). Multiple module pixellated CdTe spectroscopic X-ray detector. *IEEE Transactions on Nuclear Science, 60*, 1197–1200.
15. Meuris, A., Limousin, O., Lugiez, F., Gevin, O., Pinsard, F., Le Mer, I., Delagnes, E., Vassal, M. C., Soufflet, F., & Bocage, R. (2008). Caliste 64, an innovative CdTe hard X-ray micro-camera. *IEEE Transactions on Nuclear Science, 55*, 778–784.
16. Abbene, L., Gerardi, G., Turturici, A. A., Del Sordo, S., & Principato, F. (2013). Experimental results from Al/p-CdTe/PtX-ray detectors. *Nuclear Instruments and Methods A, 730*, 135–140.
17. Principato, F., Turturici, A. A., Gallo, M., & Abbene, L. (2013). Polarization phenomena in Al/p-CdTe/Pt X-ray detectors. *Nuclear Instruments and Methods A, 730*, 141–145.
18. Principato, F., Gerardi, G., & Abbene, L. (2012). Time-dependent current-voltage characteristics of Al/p-CdTe/Pt x-ray detectors. *Journal of Applied Physics, 112*(9), 094506.

19. Farella, I., Montagna, G., Mancini, A. M., & Cola, A. (2009). Study on instability phenomena in CdTe diode-like detectors. *IEEE Transactions on Nuclear Science, 56,* 1736–1742.
20. Abbene, L., et al. (2018). Dual-polarity pulse processing and analysis for charge-loss correction in cadmium–zinc–telluride pixel detectors. *Journal of Synchrotron Radiation, 25,* 1078–1092.
21. Buttacavoli, A., et al. (2020). Room-temperature performance of 3 mm-thick cadmium zinc telluride pixel detectors with sub-millimetre pixelization. *Journal of Synchrotron Radiation, 27,* 1180–1189.
22. Bolotnikov, A. E., Camarda, G. C., Wright, G. W., & James, R. B. (2005). Factors limiting the performance of CdZnTe detectors. *IEEE Transactions on Nuclear Science, 52,* 589–598.
23. Guerra, P., Santos, A., & Darambara, D. (2008). Development of a simplified simulation model for performance characterization of a pixellated CdZnTe multimodality imaging system. *Physics in Medicine and Biology, 52,* 589–598.
24. Koenig, T., et al. (2013). Charge summing in spectroscopic X-ray detectors with high-Z sensors. *IEEE Transactions on Nuclear Science, 60,* 4713–4718.
25. Johns, P. M., & Nino, G. C. (2019). Room temperature semiconductor detectors for nuclear security. *Journal of Applied Physics, 126,* 040902.
26. Abbene, L., et al. (2020). Recent advances in the development of high-resolution 3D cadmium–zinc–telluride drift strip detectors. *Journal of Synchrotron Radiation, 27,* 1564–1576.
27. Abbene, L., et al. (2022). Potentialities of high-resolution 3-D CZT drift strip detectors for prompt gamma-ray measurements in BNCT. *Sensors, 22*(4), 1502.
28. Howalt, O. S., et al. (2019). Evaluation of a Compton camera concept using the 3D CdZnTe drift strip detectors. *Journal of Instrumentation, 14,* C01020.
29. Butler, J. F., et al. (1992). Cd1-xZnxTe gamma ray detectors. *IEEE Transactions on Nuclear Science, 39,* 605–609.
30. Abbene, L., et al. (2020). Room-temperature X-ray response of cadmium-zinc-telluride pixel detectors grown by the vertical Bridgman technique. *Journal of Synchrotron Radiation, 27,* 319–328.
31. Zappettini, A., et al. (2011). Growth and characterization of CZT crystals by the vertical bridgman method for X-ray detector applications. *IEEE Transactions on Nuclear Science, 58*(5), 2352–2356.
32. Auricchio, N., Marchini, L., Caroli, E., Zappettini, A., Abbene, L., & Honkimaki, V. (2011). Charge transport properties in CdZnTe detectors grown by the vertical Bridgman technique. *Journal of Applied Physics, 110,* 124502.
33. Abbene, L., et al. (2017). Development of new CdZnTe detectors for room-temperature high-flux radiation measurements. *Journal of Synchrotron Radiation, 24,* 429–438.
34. Iniewski, K. (2016). CZT sensors for computed tomography: From crystal growth to image quality. *Journal of Instrumentation, 11,* C12034.
35. Chen, H., et al. (2008). Characterization of large cadmium zinc telluride crystals grown by traveling heater method. *Journal of Applied Physics, 103,* 014903.
36. Chen, H., et al. (2007). Characterization of traveling heater method (THM) grown Cd0.9Zn0.1Te crystals. *IEEE Transactions on Nuclear Science, 54*(4), 811–816.
37. Abbene, L., et al. (2016). X-ray response of CdZnTe detectors grown by the vertical Bridgman technique: Energy, temperature and high flux effects. *Nuclear Instruments and Methods A, 835,* 1–12.
38. Wang, X., et al. (2013). Further process of polarization within a pixellated CdZnTe detector under intense x-ray irradiation. *Nuclear Instruments and Methods A, 700,* 75–80.
39. Sellin, P. J., et al. (2010). Electric field distributions in CdZnTe due to reduced temperature and x-ray irradiation. *Applied Physics Letters A, 96,* 133509.
40. Bale, D. S., & Szeles, C. (2008). Nature of polarization in wide-bandgap semiconductor detectors under high-flux irradiation: Application to semi-insulating Cd1−xZnxTe. *Physical Review B, 77,* 035205.
41. Thomas, B. C., et al. (2017). Characterisation of Redlen high-flux CdZnTe. *Journal of Instrumentation, 12,* C12045.

42. Veale, M. C., et al. (2020). Characterization of the uniformity of high-flux CdZnTe material. *Sensors, 20*(10), 2747.
43. Koch-Mehrin, K. A. L., et al. (2021). Charge sharing and charge loss in high-flux capable pixelated CdZnTe detectors. *Sensors, 21*, 3260.
44. Buttacavoli, A., et al. (2022). Incomplete charge collection at inter-pixel gap in low-and high-flux cadmium zinc telluride pixel detectors. *Sensors, 22*(4), 1441.
45. Kim, J. K., et al. (2011). Charge sharing in common-grid pixelated CdZnTe detectors. *Nuclear Instruments and Methods A, 654*, 233–243.
46. Buttacavoli, A., et al. (2021). Energy recovery of multiple charge sharing events in room temperature semiconductor pixel detectors. *Sensors, 21*, 3669.
47. Buttacavoli, A., et al. (2022). Ballistic deficit pulse processing in cadmium–zinc–telluride pixel detectors for high-flux X-ray measurements. *Journal of Synchrotron Radiation, 22*, 3409.
48. Lai, X., et al. (2020). Modeling photon counting detector anode street impact on detector energy response. *IEEE Transactions on Radiation and Plasma Medical Sciences, 5*, 476. https://doi.org/10.1109/TRPMS.2020.3013245
49. Bugby, S. L., Koch-Mehrin, K. A. L., Veale, M. C., Wilson, M. D., & Lees, J. E. (2019). Energy-loss correction in charge sharing events for improved performance of pixellated compound semiconductors. *Nuclear Instruments and Methods A, 940*, 142–151.
50. Abbene, L., Gerardi, G., Raso, G., Basile, S., Brai, M., & Principato, F. (2013). Energy resolution and throughput of a new real time digital pulse processing system for x-ray and gamma ray semiconductor detectors. *Journal of Instrumentation, 8*, P07019.
51. Gerardi, G., & Abbene, L. (2014). A digital approach for real time high-rate high-resolution radiation measurements. *Nuclear Instruments and Methods A, 768*, 46–54.
52. Bettelli, M., et al. (2020). A first principle method to simulate the spectral response of CdZnTe-based X- and gamma-ray detectors. *Nuclear Instruments and Methods A, 960*, 163663.
53. Koch-Mehrin, K. A. L., et al. (2020). A spectroscopic Monte-Carlo model to simulate the response of pixelated CdTe based detectors. *Nuclear Instruments and Methods A, 976*, 164241.
54. Ballabriga, R., et al. (2020). Photon counting detectors for X-ray imaging with emphasis on CT. *IEEE Transactions on Radiation and Plasma Medical Sciences, 5*, 422. https://doi.org/10.1109/TRPMS.2020.3002949
55. Abbene, L., et al. (2018). Digital fast pulse shape and height analysis on cadmium–zinc–telluride arrays for high-flux energy-resolved X-ray imaging. *Journal of Synchrotron Radiation, 25*, 257–271.

Effect of Electron Irradiation on Spectroscopic Properties of Schottky Barrier CdTe Semiconductor Detectors

Katarína Sedlačková, Bohumír Zaťko, Andrea Šagátová, and Vladimír Nečas

1 Introduction

High-Z semiconductors are due to their high absorption coefficient very suitable candidates for detection of X- and gamma-rays of higher energies, where the low-density materials, e.g., silicon, exhibit only poor detection efficiency. Although the radiation detector technologies based on materials like high purity germanium (HPGe) are providing excellent energy resolution, they are for many applications unattractive especially due to the requirements on large cryogenic cooling system and at least also because of their limited spatial resolution. Detector technologies based on compound semiconductors of high-density materials meet the requirements of a high energy and spatial resolution and a capability of routine operation at room temperature due to their high electrical resistance. The most frequently used compound semiconductors are binary materials such as gallium arsenide (GaAs), cadmium telluride (CdTe) or indium phosphide (InP); or ternary materials, e.g., cadmium zinc telluride (CdZnTe). The term compound semiconductor encompasses a wide range of materials, most of which crystallize in either the zinc-blende, wurtzite, or rocksalt crystal structures [1]. They are generally derived from elements in groups II to VI of the periodic table. Modifying the composition of a compound semiconductor changes its average lattice constant and consequently the band-gap energy, which is dependent on the inter-atomic distance and electronegativity of the constituent elements. Additional benefit of compound semiconductors as compared to their elemental counterparts is therefore the possibility to grow materials with a

K. Sedlačková (✉) · B. Zaťko · A. Šagátová · V. Nečas
Slovak University of Technology in Bratislava, Institute of Nuclear and Physical Engineering,
Bratislava, Slovak Republic
e-mail: katarina.sedlackova@stuba.sk

wide range of physical properties (e.g., atomic number, density, band-gap) enabling to engineer a material suitable to almost any application [1, 2].

The choice of which compound to use for a specific application depends to a large extent on the energy range of interest. For a given compound having N atoms per unit volume and Z being its effective atomic number, the interaction cross-section varies as NZ^5 for photoelectric effect (dominates in the energy region up to 200 keV) and NZ^2 for Compton effect (up to a few MeV), whereby only the photoeffect leads to the total absorption of the incident photon energy. Thus, for an effective absorption of X- and gamma-ray photons, a material with highest possible Z is required to increase the energy range for the total energy absorption.

Atomic number of the elements constituting a detector influences also one of the most important spectrometric parameters – an energy resolution. Energy resolution of a detection system is given by the statistics of carrier generation (Fano noise or broadening), by the electronic noise (detector leakage current and preamplifier noise) and by the noise due to incomplete charge collection due to carrier trapping. Fano noise is influenced mainly by the average energy consumed to create an electron-hole pair. For compound radiation detectors, the noise due to incomplete charge collection is, however, more important as compared to Fano and electronic noise (except of the thin detectors) [1].

Despite the abovementioned advantageous properties of compound semiconductor detectors, there are still many significant technological challenges to make them ready for mass distribution and/or to customize their properties to a specific application within the scientific, medical, or security sectors. Compound semiconductors are plagued especially by materials problems caused by high defect densities and impurities. The impurities are acting as trapping centers for charge carries and are negatively affecting the charge transport properties of compound semiconductors. Consequently, the charge carrier suffers from either poor mobility μ or carrier lifetime τ. The mobility-lifetime product ($\mu\tau$) of compound semiconductors is of the order of 10^{-4} for electrons and 10^{-5} for holes and are getting worse for higher Z materials. For comparison, in case of elemental semiconductors, the product $\mu\tau$ reaches the order of unity for both electrons and holes.

Another issue affecting the detector performance is the quality of metal-semiconductor contacts, which can be either ohmic or Schottky (p-n junction). Detectors with ohmic contacts obeying the Ohm's law operate at low bias voltages (tens of volts) and have good time stability of functional parameters (detection efficiency and energy resolution). Their energy resolution is, however, poorer due to incomplete collection of generated charge carries. Higher bias voltages cannot be applied because of an increase in the leakage (dark) current. The applied bias voltage can be increased up to hundreds of volts when replacing one ohmic contact with a contact with Schottky barrier or p-n junction acting like a rectifying diode, which decreases in turn the leakage current [3]. The fabrication of stable and uniform contacts in compound semiconductor detectors is also a problem due to the limited choice of suitable materials.

2 Cadmium Telluride Detectors

Cadmium telluride (CdTe) is a II-IV compound semiconductor with a zinc-blende crystal structure. It has been studied as an X-ray and gamma-ray detector material since the 1960s [2]. Radiation detectors based on CdTe are being successfully used since decades in many fields of industry covering the energy range from a few keV to MeV [4]. Particularly medical imaging like nuclear cameras or positron emission tomography systems, and various astrophysical applications, e.g., high-performance spectrometers for the universe exploration, prosper from their excellent efficiency for the X-ray and gamma-ray detection [2, 5, 6]. This attractive property results from a high absorption coefficient of CdTe compound due to its high density of 5.85 g.cm^{-3} and high atomic number of the constituent elements ($Z_{Cd} = 48$, $Z_{Te} = 52$). Additionally, CdTe possess a relatively low average energy consumed to create an electron-hole pair of about 4.43 eV [1]. Sufficiently high band-gap energy of CdTe of about 1.5 eV implying high room temperature resistivity in the interval of 0.5–5 $\times$ 10^9 Ωcm allows larger biases to be applied (resulting in faster drift velocities and deeper depletion depths) and are a great advantage for their room temperature operation. The mentioned favorable characteristics induced extensive international effort invested into the development of various CdTe-based detection systems, with the aim to improve their overall performance matching a specific application or environment.

One of the former disadvantages of CdTe-based detectors was the limitation of their energy resolution due to insufficient structural quality of CdTe bulk crystals [7]. These material and fabrication difficulties have been overcome at the beginning of the 1990s, and commercial availability of high-quality crystals has improved. Presence of defects and impurities in the crystals acting as trapping centers results, however, still in poor charge carrier transport properties which are typical for CdTe materials [2]. In CdTe detectors with Schottky contacts, application of high reverse bias voltages is therefore required to increase charge collection and to avoid trapping of nonequilibrium carriers created during photon absorption. Higher bias voltages ensure excess of charge carries either by tunneling mechanism or by injecting minority charge carriers from the imperfect ohmic contact [7].

Material quality in combination with the nature of the electrical contacts plays a key role in another since decades ago known critical issue of the CdTe detectors, the so-called polarization effect affecting especially CdTe detectors with Schottky contacts [4, 7–16]. It causes degradation of detector spectrometric characteristics when operating longer under applied reverse bias voltage, thus hindering the reliability of long-time measurements.

The reliability of radiation detectors can be also strongly affected by a damage induced by high-energy particles. As their onset in radiation harsh environments is often foreseen (e.g., various astrophysical applications), radiation hardness of semiconductor detectors accounts to important attributes for their practical operation. Pending radiation damage can cause an increase of the leakage current, degradation of energy resolution, and a shift of the peak position as a consequence of reduced charge collection efficiency [17]. In our study, we have focused on the radiation damage of CdTe Schottky barrier detector induced by high-energy electrons. We have also analyzed the effect electron irradiation on the detector time stability.

3 Polarization Effect

The time instability of CdTe-based detectors known as polarization effect remains in focus for the scientific community since many years due to a limitation of the practical use of a detector. Polarization effect causes progressive time degradation of detector performances after applying bias voltage and limits their application at room temperature. The most typically we observe worsening of detection efficiency manifested by progressive decrease of photopeak amplitude, deterioration of energy resolution (an increase of the Full Width at Half Maximum, FWHM), and a shift of the photopeak position toward lower energies with time after applying bias [4, 12, 15, 16]. The most sensitive to polarization effect are the Schottky-diode detectors, even though it has been reported that CdTe detectors based on p-n junction diodes also show some time instability of electrical and spectral properties attributed to the polarization phenomenon [4].

The origin of polarization effect is still debated. The most plausible explanation is the non-uniform electric field formation in the CdTe bulk due to the accumulation of negative charge carriers at deep acceptor levels [2, 4, 7, 13, 18]. Though polarization is known as a bulk phenomenon, there are indications that the rate of polarization is very dependent on the electrode interfaces and on the surface preparation technique [9, 16]. Toyama et al. [19] proposed a modification of charge accumulation model, which is based on the assumption that deep acceptor levels in a region close to the Schottky contact are already ionized before applying a voltage. If a reverse bias is applied, deep acceptor levels begin to be occupied, and negative space charge is accumulated. The electric-field strength at the Schottky contact increases with time which causes a deterioration of energy resolution. On the other hand, the electric field at the cathode decreases with time suggesting that the width of the depletion layer begins to diminish and becomes less than the thickness of the bulk at a certain time, which corresponds with the time that the photopeak position begins to shift. So, it is considered that the shift of photopeak position in radiation measurements is due to the decrease in depletion width. Meuris et al. [13] have emphasized that another thermal effect has to be taken into account to explain spectral response of Schottky CdTe detectors. She showed that the polarization phenomenon is even more advanced before applying a voltage as the temperature is high.

The voltage-induced polarization is reversible and can be reduced or even avoided by either changing the contact potential barrier [14] or by periodically switching off the bias voltage (using periodical impulse supply mode [15]). Niraula et al. [9] and Meuris et al. [13] reported that better detector stability can be obtained with a moderate cooling or by applying a higher electric field, whereby the effect of cooling on the stability was found to be more effective than applying higher electric fields. Yamazato et al. [18] presented another improvement in the polarization of the CdTe radiation detectors by the sulfur treatment (5 times longer operation time as compared to an untreated detector).

In general, the term polarization can be used to denote any externally induced time instability of detector performances. This can be not only applying of bias voltage to the detector but also irradiating it by intense X-ray or gamma-ray sources with high photon fluency [20]. In this case, the recovery of detector parameters after bias resetting is only partial.

The above-described time instability of a detector can be quantified by a parameter characterizing the polarization onset, the so-called polarization time t_P, corresponding to the moment when the electric field at the cathode drops down to zero [12]. In the spectroscopic measurements, the polarization time can be derived from the analysis of a photopeak characteristic (peak position, FWHM, or peak area) evaluated as a function of time. The polarization time depends strongly on the reverse bias voltage height and on the temperature. In our study, we have followed its dependency on the doses delivered to the detectors by high-energy electron irradiation. Although the phase before polarization time is referred to as a stable period, the model of electric field presented in [13] predicts a peak shift as soon as the detector is biased. Therefore, we have introduced the second parameter reflecting the spectroscopic parameter stability in the stable operational phase before polarization onset, which is the slope of a spectrometric parameter time dependency before polarization onset (e.g., the peak position in our case). The slope of this dependency should be for a reliable operation as low as possible. However, it can be also influenced by some external factors, as temperature changes or high-energy particle irradiation.

4 Radiation Hardness of CdTe Detectors

Radiation detectors are exposed during their operation to substantial fluxes of ionizing radiation leading to formation of radiation-induced defects. Their accumulation causes degradation of detector parameters and limits the radiation durability of detectors. The intensity of detector radiation damage depends mainly on the detector material, on the type of radiation, on the dose applied, and on the particle fluence. For example, charged particles seem to be much more effective than uncharged particles at comparable irradiation doses in reducing the spectroscopic capabilities of detectors [21].

There are many studies dealing with radiation resistance of CdTe-based detectors against protons, X-rays, and gamma-rays (Table 1). The degradation in CdTe detectors under a gamma-irradiation begins at substantially higher doses than for the Si and Ge semiconductor detectors [22]. As mentioned above, CdTe-based detectors might be attractive for different kinds of astrophysical applications, where the devices are constantly subjected to electron bombardment of high-energy electrons (MeV). Nevertheless, studies using high-energy electrons are quite sporadic. In Ref.

Table 1 List of radiation resistance studies of CdTe-based semiconductor detectors

Radiation type	Energy	Radiation dose/fluence	Ref.
Protons	2 MeV	Up to 10^{10} p/cm^2	[17]
	3 MeV	Up to 10^{10} p/cm^2	[23]
	22 MeV	Up to 10^{12} p/cm^2	[24]
	200 MeV	Up to 10^{10}–10^{11} p/cm^2	[25]
Photons	6 MeV	7–25 kGy	[26]
	1.17 MeV; 1.33 MeV (^{60}Co)	10–50 kGy	[21, 27]
Neutrons	Thermal	10^{10}–10^{14} n/cm^2	[21, 27]
	Fast	10^{11}–10^{13} n/cm^2	[21, 27]
	Up to 1 MeV	Up to 10^{15} n/cm^2	[28]
Electrons	7.5–12 MeV, mean 9 MeV	2 and 12 kGy	[21]

[21], A. Cavalini et al. presented results of CdTe:Cl detectors irradiation with an electron beam characterized by an energy spectrum ranging from 7.5 to 12 MeV. The delivered dose varied between 2 and 12 kGy. It was found out that electron irradiation has a dramatic effect on the CdTe:Cl detectors: a 2 kGy dose completely degraded their spectroscopic capabilities. With this respect, we have started our study to explore the mechanisms of electron radiation damage of CdTe Schottky barrier detectors with the aim to follow the behavior of their spectrometric properties and to assess the maximum tolerable dose at which their operation is still reliable. Moreover, there is no study inspecting the effect of high-energy electrons on the time instability of the Schottky barrier CdTe detectors to our knowledge. Therefore, we have focused also on the analysis of the polarization onset and on the detector stability in the phase before polarization and have followed detector spectrometric parameters with respect to the applied dose height.

For comparison, the radiation hardness of another high Z-compound semi-conductor detectors based on semi-insulating GaAs can be found in [29]. The presented analysis points at different damaging properties of high-energy electrons as compared to high-energy gamma-rays or neutrons. According to the obtained results, the studied GaAs sensors were still functional after applying a dose of 1140 kGy of about 1 MeV photons, 104 kGy of 5 MeV electrons, but only up to 0.576 kGy of fast ($\sim$2–30 MeV) neutrons. Hence, it was confirmed that electron irradiation is less damaging than neutron irradiation of similar energy and dose, but has higher degradation effect than gamma-rays of MeV energies [30]. The continued study on radiation hardness of GaAs sensors against electrons presented in [31] showed that in spite of systematic deterioration of their spectrometric parameters with the applied dose, the detectors proved to be functional up to a cumulative dose of 1000 kGy.

5 Experimental Details

5.1 Detector Fabrication

The studied CdTe Schottky barrier detectors were based on the high-quality single crystal of CdTe manufactured by Acrorad Co., Ltd., Japan. The bare chips were of size $4 \times 4 \times 1$ mm^3 and had an indium-titanium multilayer Schottky barrier contact and a platinum contact (Fig. 1a). The platinum contact was fabricated by the electroless plating (thickness of ca 20 nm), and the indium and titanium contact were formed by the vacuum evaporation (thickness of ca 300 nm and 30 nm, respectively) [32]. The use of indium on CdTe material as hole blocking contact results in high hole barrier height and thus allows very high average electric fields to be applied without considerable current increase [16]. Excellent spectral characteristics of the sensors have been achieved due to production of homogeneous large-volume single crystals by Acrorad and an improved technique for depositing the indium Schottky metal contact.

According to the enclosed specification, the sensors exhibited the FWHM (full width at half maximum) energy resolution of 3.2–3.3 keV at 59.5 keV and bias 700 V and the leakage current of 0.94–1.32 nA at 500 V and 1.28–2.07 nA at 700 V, respectively [33]. The chips were wire bonded and encased at the Slovak Academy of Sciences in Bratislava. A photograph of a fabricated detector is presented in Fig. 1b. CdTe Schottky detectors with indium contacts are unfortunately not stable at room temperature when biased and their charge collection properties degrade with time through the mechanism described above as declared also by the chips manufacturer [32].

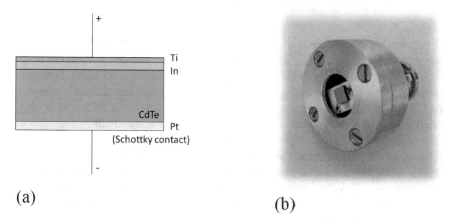

(a)

(b)

Fig. 1 (**a**) The schematic structure of the used detector chip from Acrorad Co.; (**b**) Encased CdTe Schottky barrier detector

6 Electron Irradiation

CdTe detectors were irradiated with a scanning electron pulsed beam with the mean energy of 5 MeV at the accelerator exit at the linear accelerator UELR 5-1S (University Centre of Electron Accelerators, Trenčín). Accelerator and beam parameters during irradiation were as follows: single-pulse duration 3.5 μs, scanning frequency 0.25 Hz, beam repetition rate 10 Hz, beam scanning width 40 cm, beam diameter at the sample 8 cm, and average beam current 5 μA. Detectors were placed on the 1-cm-thick aluminum board and irradiated from the detector Schottky contact side. The distance between the detectors and the foil of the accelerator vacuum exit window was 55 cm. The absorbed doses were measured on detector surfaces using B3 radiochromic films, evaluated by Spectrophotometer GENESYS20 and verified with RISO polystyrene calorimeters.

Studied detector samples were irradiated simultaneously to avoid the variations in applied doses. In the first irradiation trial, based on our experience with radiation resistant SiC and GaAs detectors tolerating hundreds of kilograys [31, 34], the initial cumulative dose of 10 kGy was applied to the studied CdTe detectors. After irradiation experiment, the detectors were almost destroyed showing unacceptable spectrometric properties. Based on these results, we have started a new experiment with the lowest dose delivered of 0.5 kGy proceeding stepwise to the total cumulative dose of 2.25 kGy. The upper limit of the applied dose was governed by the obtained results showing the immediate onset of polarization after applying the bias voltage at the highest applied dose. After each irradiation step, the spectroscopic and electrical measurements (IV-characteristics) were carried out.

7 Spectroscopic Measurements

Spectroscopic performance of detectors over time has been characterized with spectroscopic measurements using [241]Am radioisotope source for different doses delivered to detectors by high-energy electron beam (up to 2.25 kGy). Detectors were connected to a standard spectrometric chain with Amptek readout electronics (A250CF CoolFET charge sensitive preamplifier) and Amptek PX5 device comprising a digital pulse processor (DPP), a multichannel analyzer (MCA) and power supply connected to a computer running data acquisition and control software DPPMCA. All measurements were carried out at room temperature.

Radiation measurements were performed in various conditions of applied voltages (300 V, 400 V, and 500 V). The upper value of the bias voltage of 500 V was found to be sufficiently high to obtain excellent spectrometric properties of the studied detectors, and it was also determined by the technical limitations of the used spectrometric chain.

Correct operation of CdTe detector over time are conveniently evaluated by following the spectral parameters of the 59.5-keV photopeak, like the peak position

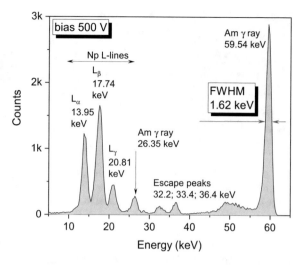

Fig. 2 Energy spectra of ^{241}Am source acquired using a CdTe Schottky barrier detector before electron irradiation showing the best achieved energy resolution of 1.62 keV at 59.54 keV (bias 500 V)

(giving information about the charge collection efficiency), the peak width (related to energy resolution in the term of FWHM), and the peak area (pointing at possible changes of the detection efficiency). To analyze the time stability of the chosen spectral parameters, the function of DPPMCA software enabling the automatic collection of a user defined spectra sequence (pre-set acquisition time and number of spectra) was applied during spectra acquiring process. The spectra were collected in 30 s or in 2 min intervals, depending on the rate of their deterioration. The software tool SeGaSA [35] was used for automatic determination of the followed photopeak parameters for each spectrum in the sequence, namely, the peak position, the peak width and height, and the peak area [36].

The energy resolution of CdTe Schottky detectors depends significantly on the applied bias, whereby the higher the bias applied the better the energy resolution. Figure 2 shows the spectral response of a sample detector to ^{241}Am gamma-source demonstrating the best achieved energy resolution of 1.62 keV (2.7%) at 59.5 keV (bias of 500 V) before electron irradiation.

8 Simulations of Electron Irradiation

Before irradiation experiment, simulation of 5 MeV electron transport was performed in order to determine: (a) the energy of electrons impinging the detector surface after passing the distance between accelerator vacuum exit window and the detector in air and (b) the real total dose delivered to the detector volume with respect to the dose measured on the sample surface. Simulations of electron transport were realized using CASINO v2.51 program which is a single scattering Monte Carlo simulation of electron trajectory in solid [37] and using ModePEB program [38].

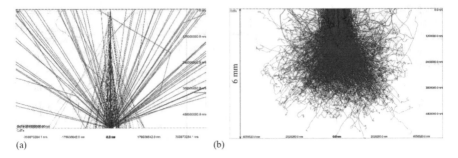

Fig. 3 Trajectories of 5 MeV electrons in air impinging 1-mm-thick CdTe detector at a distance of 55 cm (**a**); trajectories of 4.86 MeV electrons in CdTe bulk material demonstrating mean electron range of 3.14 mm (**b**), simulated using CASINO v2.51 program

Figure 3a illustrates 5 MeV electron trajectories traveling through 55 cm of air and reaching the CdTe detector surface (blue lines) simulated using CASINO program. Red lines in the figure represent the trajectories of backscattered electrons. We have calculated that after passing 55 cm of air, the mean energy of electrons decreases to 4.86 MeV. Figure 3b shows the simulation of 4.86 MeV electron paths in CdTe bulk material used to calculate their range. The range has been determined as the mean value of the individual projected ranges and reached the value of 3.14 mm, which is apparently in accordance with electron paths showed in Fig. 3b.

Figure 4 presents the calculated decrease of the mean electron energy as a function of the material depth. As visible from the graph, the mean energy of electrons leaving 1-mm-thick detector material is about 3.98 MeV. The energy losses calculated from the energy-depth curve (Fig. 4) show expected increasing tendency and rise from about 0.71 keV/μm (on the top of the detector) to about 1.04 keV/μm (bottom of the detector). Figure 5 shows the dose depth distribution induced by 4.86 MeV electrons in CdTe determined using a complementary ModePEB program. The dose is increasing from the surface, reaches its buildup maximal value of about 133% (with respect to the surface dose) in the depth of about 0.47 mm, and starts decreasing toward the bottom of the detector material. The electron energy deposition in the top Schottky contact was neglected in the calculations.

It should be pointed out that the electron-beam energy-loss profile does not coincide 1:1 with the energy-deposition profile. There are several reasons for this effect. A non-negligible part of the kinetic energy lost by the primary electron beam is converted to bremsstrahlung. At 5 MeV and atomic number ~50, the ratio between the radiative losses and ionizing losses is about 0.3. The final dose profile is therefore a superposition of the bremsstrahlung dose profile and the primary electron-beam ionization dose profile. The bremsstrahlung photons may deposit their energy either in different location of the detector or may even escape from the detector without being absorbed in the detector volume at all. The delta electrons produced by ionization in the sub-surface region may also be scattered out of the

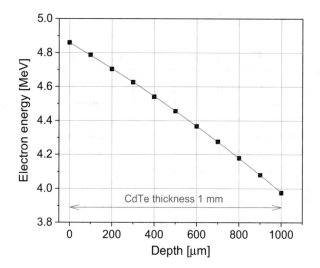

Fig. 4 Mean energy of the 4.86 MeV electron beam as a function of depth in CdTe simulated using CASINO v2.51 program

Fig. 5 The dose depth distribution induced by 4.86 MeV electrons in CdTe (density of 5.859 g/cm^3) according to simulation in ModePEB

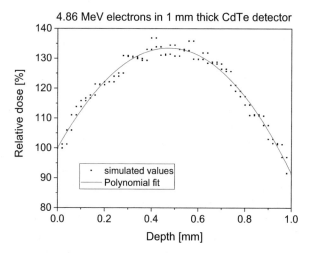

detector taking away some part of the energy lost by the primary electron beam. That is why in general the dose profile differs from the energy-loss profile and exhibits the typical buildup maximum below the surface.

The simulation results point at the fact that the dose measured during the irradiation on the top of the detector underestimates slightly the real dose deposited in the whole detector volume. According to the simulation results, the real dose delivered reaches about 120% with respect to the dose measured on the sample surface.

9   Results and Discussion

Figure 6 presents the ^{241}Am spectra measured stepwise with an interval of 120 s at different reverse bias of 300 V, 400 V, and 500 V demonstrating the polarization effect before (left side panels) and after (right side panels) electron irradiation with the highest applied dose of 2.25 kGy for a sample CdTe Schottky barrier detector. The spectra show a progressive degradation after applying bias – a shift of the 59.5 keV-peak position toward lower energies and a decrease of the photopeak height and area. Comparing detector spectra before irradiation (left-side panel in Fig. 6), better spectral characteristics, and slower degradation with time was observed for the higher bias voltages applied. After electron irradiation (right-side panel in Fig. 6), the spectrometry of the detector gets worse rapidly, and, additionally, the time stability of the response is decreased (faster spectra degradation with time).

The above-described tendencies can be quantified by spectral parameters derived from the 59.5 keV-photopeak, which are presented in Fig. 7 before and after irradiation. The figure shows the time evolution of four parameters (peak position, FWHM, net peak area, and peak height) for measurements at different reverse bias voltages (300 V, 400 V, and 500 V) before irradiation (full graph points) and after irradiation with the highest dose of 2.25 kGy (empty graph points). Concerning the unirradiated sample, one can follow that the spectral parameters begin to degrade at a certain time, which becomes longer as the bias voltage increases. The dependence of the peak centroid (top-left corner of Fig. 7) on the time of biasing can be divided into two regions. The first region is characteristic by a gentle linear decrease, and it is followed by a more rapid drop, whereby different rates of the pertinent tendencies carry information about polarization onset. The data have been accordingly fitted by two linear approximations whose intercept can be denoted as the polarization time, t_P. Using described evaluation procedure, we obtained polarization times for unirradiated detector samples of about 49 min, 60 min, and 101 min for the measurements at 300 V, 400 V, and 500 V, respectively. Increasing polarization time with bias voltage confirms the improvement of the detector time stability. Polarization time can be derived from other spectral parameters reported in Fig. 7 using the same procedure as described above, expecting to provide similar results. In the top-right corner of Fig. 7, the time evolution of the energy resolution in terms of the FWHM is plotted. This spectral parameter remains relative stable until the polarization onset and, after that, increases rapidly. The onset of polarization is equally earlier at lower reverse bias voltages. It has to be taken into account that the poorer resolved and the more asymmetric peak, the less reliable the determination of the FWHM parameter becomes. The bottom-left corner of Fig. 7 presents the net area of the photopeak, and the bottom-right corner shows the last parameter evaluated – the height of the photopeak. Both characteristics depict analogical tendencies as described formerly, a stable period followed by a sudden break getting on later for higher bias voltages applied.

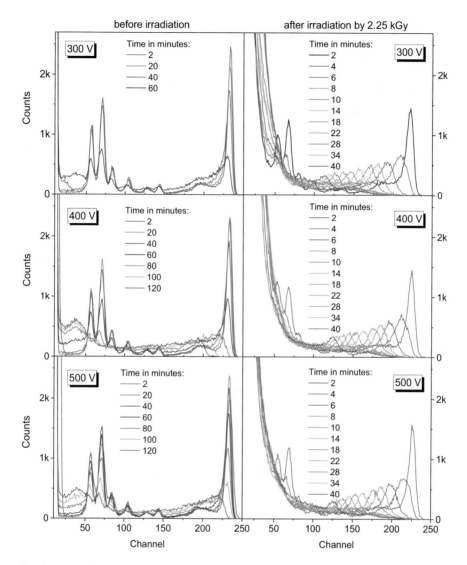

Fig. 6 Polarization effect for CdTe detector before (left side panel) and after (right side panel) applied dose of 2.25 kGy

In Fig. 7 we can also observe the time evolution of the spectrometric characteristics after irradiation with the highest cumulative dose of 2.25 kGy (empty points in graphs). In comparison with responses of the unirradiated detector, the sequentially acquired spectra of a detector irradiated by 2.25 kGy demonstrate the lack of a stable period and a dramatic shift of the peak position, a decrease of both the net peak area and the peak amplitude, and a rapid increase of the FWHM directly after applying the bias voltage. These observations indicate a significant effect of electron

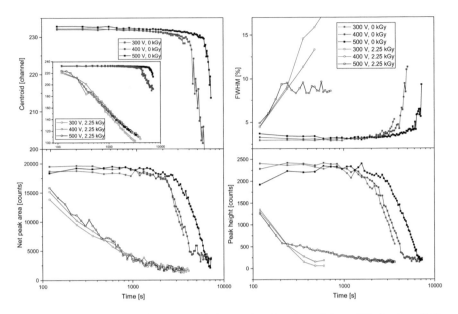

Fig. 7 Spectral parameters of the 59.5 keV-photopeak as functions of time of biasing for a CdTe detector before and after electron irradiation with a dose of 2.25 kGy at different bias voltages (300 V, 400 V, and 500 V)

irradiation on the polarization onset, which apparently follows immediately after applying the bias voltages when the samples are damaged by electron irradiation with a cumulative dose of 2.25 kGy.

To analyze the effect of electron irradiation on detector spectrometric performance with relation to the dose delivered, we have chosen the spectra collected at the reverse bias voltage of 500 V due to the best spectrometry, as shown before. Figure 8 presents the time evolution of spectral parameters derived from the measured detector responses after each single irradiation step delivering successively a dose of 0.5; 0.75; 1.0; 1.5; 1.75; and 2.25 kGy, respectively. As obvious from the graphical interpretation of the individual spectral parameters, their behavior in time shows similar tendencies of worsening as described earlier, although their dramatic degradation after polarization onset does not follow a linear tendency, but rather an exponential decrease (note the semilogarithmic scale in Fig. 8). Already the lowest dose of 0.5 kGy applied to detectors induced substantially earlier onset of polarization, and toward the highest obtained dose of 2.25 kGy, the stable period is approaching zero.

Figure 9 presents the polarization time derived from the peak position time dependencies as a function of the dose delivered (the spectra collected at the reverse bias voltage of 500 V). The graphical interpretation clearly shows a rapid drop of the polarization time with increasing dose. The red line in the plot represents the proposed exponential fit of the data, which describes the dependency of the

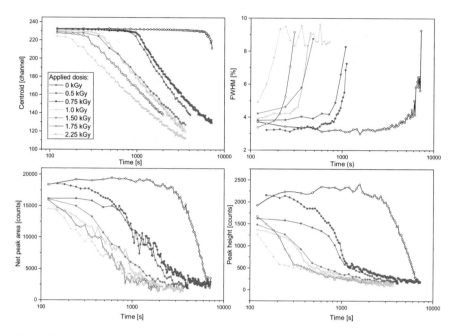

Fig. 8 Spectral parameters of the 59.5 keV-photopeak as functions of time of biasing for unirradiated CdTe detector and detector irradiated stepwise up to a cumulative dose of 2.25 kGy acquired at a bias voltage of 500 V

polarization time, t_p (in minutes), on the dose delivered, D (in Grays) as follows: $t_p = t_0 \exp(-K \cdot D)$, where the parameters are at the size of $t_0 = 98$ min and $K = 3.6 \ kGy^{-1}$. Discussion of some physical aspects concerning the polarization time dose dependency can be found in [39].

Except for the polarization time, we have also studied the effect of electron irradiation on detector spectrometry in the stable operational phase before polarization onset. With this aim, we have evaluated the slope of the peak position time shift as a function of the dose delivered. A gentle time shift of the peak position before polarization observed for an unirradiated detector tends to increase rapidly (in its absolute value) after electron irradiation. The details on this study can be found in [40].

The electric parameters of the detectors are also significant indicators of radiation induced changes and can be controlled by current-voltage characteristics measurements. In our case, the measurements performed after each irradiation step proved functionality of the detectors and showed slight decrease of the dark current down to about 1 nA at a reverse bias voltage of 500 V after first irradiation step [33]. Increasing the applied dose, we have observed fluctuations of the dark current of around 1 nA. The dark current changes of this order might be partly caused also by the lack of temperature stabilization during experiments.

Fig. 9 Polarization time t_P plotted as a function of the applied dose for a sample detector at a bias voltage of 500 V

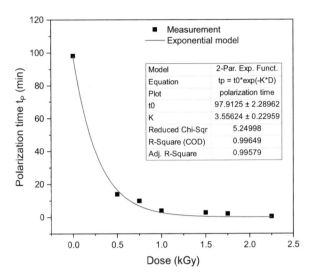

10 Summary

The use materials with a high proton number provides a great advantage in the design of semiconductor detectors, especially due to the possibility to achieve a high detection efficiency of high-energy X-rays and gamma-rays detection. Studied Schottky barrier detectors manufactured from cadmium telluride chips (Acrorad) demonstrated excellent spectrometry especially at higher reverse bias voltages applied. The best energy resolution of 1.62 keV was acquired at RT for [241]Am radioisotope source generated gamma-rays of 59.54 keV at a bias voltage of 500 V. Detector irradiation with 5 MeV-energy electrons caused, however, rapid degradation of detector spectrometric properties already when irradiated by a relatively low dose of 0.5 kGy. A shift of the photopeak, an increase in FWHM, and a decrease in peak height and net area were observed, which progressed with increasing applied dose up to 2.25 kGy, where the degradation of detectors hindered their reliable operation.

Since we know the disadvantageous time stability typical for Schottky barrier CdTe detectors caused by the polarization phenomenon, the effect of the electron irradiation on the detector time stability was further studied. The time dependence of the evaluated spectral parameters points to the fact that the duration of the detectors stable phase is highly sensitive to the applied radiation dose height. In the examined dose range from 0.5 kGy up to 2.25 kGy, the results showed a faster polarization onset for increasing total cumulative dose (an exponential decrease of the polarization time with the dose). The observed degradation of the detector spectrometric characteristics, such as the charge collection and detection efficiency and the energy resolution, indicates that electron irradiation might induce changes of the depletion layer width and the electric-field strength at the Schottky contact.

The electron irradiation affects, however, not only the polarization onset but also the spectroscopic performance of detectors during the operational phase before the break. Although detector spectrometry tends to degrade only insignificantly with time in this relatively stable region, the measurements indicated decreased time stability after electron irradiation.

Acknowledgement This work was partially supported by grants of the Slovak Research and Development Agency No. APVV-18-0243 and of the Scientific Grant Agency of the Ministry of Education, Science, Research and Sport of the Slovak Republic No. 2/0084/20.

References

1. Owens, A., & Peacock, A. (2004). Compound semiconductor radiation detectors. *Nuclear Instruments and Methods in Physics Research, Section A: Accelerators, Spectrometers, Detectors and Associated Equipment, 531*(1–2), 18–37.
2. del Sordo, S., Abbene, L., Caroli, E., Mancini, A. M., Zappettini, A., & Ubertini, P. (2009). Progress in the development of CdTe and CdZnTe semiconductor radiation detectors for astrophysical and medical applications. *Sensors, 9*(5), 3491–3526.
3. Gnatyuk, V., Maslyanchuk, O., Solovan, M., Brus, V., & Aoki, T. (2021). CdTe X/-ray detectors with different contact materials. *Sensors, 21*(3518), 1–20.
4. Gnatyuk, V. A., Vlasenko, O. I., Aoki, T., & Koike, A. (2016). Characteristics and stability of diode type CdTe-based x-ray and gamma-ray detectors. *2014 IEEE Nuclear Science Symposium and Medical Imaging Conference*, NSS/MIC 2014 7431267
5. Takahashi, T., et al. (2001). High-resolution CdTe detector and applications to imaging devices. *IEEE Transactions on Nuclear Science, 48*(3 I), 287–291.
6. Szeles, C. (2004). CdZnTe and CdTe materials for X-ray and gamma ray radiation detector applications. *Physica Status Solidi (B): Basic Research, 241*(3), 783–790.
7. Maslyanchuk, O., et al. (2021). Charge transport features of CdTe-based X- and γ-ray detectors with Ti and TiOx Schottky contacts. *Nuclear Instruments and Methods in Physics Research, Section A: Accelerators, Spectrometers, Detectors and Associated Equipment, 988*, 164920.
8. Malm, H. L., & Martinit, M. (1974). Polarization phenomena in CdTe nuclear radiation detectors. *IEEE Transactions on Nuclear Science, 21*, 322–330.
9. Niraula, M., Nakamura, A., Aoki, T., Tomita, Y., & Hatanaka, Y. (2002). Stability issues of high-energy resolution diode type CdTe nuclear radiation detectors in a long-term operation. *Nuclear Instruments and Methods in Physics Research, Section A: Accelerators, Spectrometers, Detectors and Associated Equipment, 491*(1–2), 168–175.
10. Takahashi, T., et al. (1999). High-resolution Schottky CdTe diode for hard X-ray and gamma-ray astronomy. *Nuclear Instruments and Methods in Physics Research, Section A: Accelerators, Spectrometers, Detectors and Associated Equipment, 436*(1–2), 111–119.
11. Matsumoto, C., Takahashi, T., Takizawa, K., Ohno, R., Ozaki, T., & Mori, K. (1998). Performance of a new Schottky CdTe detector for hard X-ray spectroscopy. *IEEE Transactions on Nuclear Science, 45*(3), 428–432.
12. Principato, F., Turturici, A. A., Gallo, M., & Abbene, L. (2013). Polarization phenomena in Al/p-CdTe/Pt X-ray detectors. *Nuclear Instruments and Methods in Physics Research, Section A: Accelerators, Spectrometers, Detectors and Associated Equipment, 730*, 141–145.
13. Meuris, A., Limousin, O., & Blondel, C. (2011). Characterization of polarization phenomenon in Al-Schottky CdTe detectors using a spectroscopic analysis method. *Nuclear Instruments and Methods in Physics Research, Section A: Accelerators, Spectrometers, Detectors and Associated Equipment, 654*(1), 293–299.

14. Farella, I., Montagna, G., Mancini, A. M., & Cola, A. (2009). Study on instability phenomena in CdTe diode-like detectors. *IEEE Transactions on Nuclear Science, 56*(4), 1736–1742.
15. Petukhov, Y., & Krop, W. (2008). The research of polarization in CdTe P-I-N detectors of nuclear radiation. *2008 IEEE Nuclear Science Symposium Conference Record* (pp. 263–268).
16. Cola, A., & Farella, I. (2009). The polarization mechanism in CdTe Schottky detectors. *Applied Physics Letters, 94*(10), 1–5.
17. Zanarini, M., et al. (2004). Radiation damage induced by 2 MeV protons in CdTe and CdZnTe semiconductor detectors. *Nuclear Instruments and Methods in Physics Research, Section B: Beam Interactions with Materials and Atoms, 213*, 315–320.
18. Yamazato, M., et al. (2006). Improvement in y-ray detection quality of Al/CdTe/Pt Schottky-type radiation detector by sulfur treatment. *Japanese Journal of Applied Physics, Part 2: Letters, 45*(46–50), L1263.
19. Toyama, H., Higa, A., Yamazato, M., Maehama, T., Ohno, R., & Toguchi, M. (2006). Quantitative analysis of polarization phenomena in CdTe radiation detectors. *Japanese Journal of Applied Physics, Part 1: Regular Papers & Short Notes Revised Paper, 45*(11), 8842–8847.
20. Bale, D. S., & Szeles, C. (2008). Nature of polarization in wide-bandgap semiconductor detectors under high-flux irradiation: Application to semi-insulating Cd1-x Znx Te. *Physical Review B: Condensed Matter and Materials Physics, 77*(3), 1–16.
21. Cavallini, A., et al. (2002). Radiation effects on II-VI compound-based detectors. *Nuclear Instruments and Methods in Physics Research A, 476*, 770–778.
22. Nasieka, I., et al. (2019). Increased radiation hardness of detector-grade Cd0.96Zn0.04Te crystals by doping with In and Ge. *Radiation Physics and Chemistry, 165*(April), 108448.
23. Pastuović, Ž., & Jakšić, M. (2001). Frontal IBICC study of the induced proton radiation damage in CdTe detectors. *Nuclear Instruments and Methods in Physics Research, Section B: Beam Interactions with Materials and Atoms, 181*(1–4), 344–348.
24. Ahoranta, J., et al. (2009). Radiation hardness studies of CdTe and HgI2 for the SIXS particle detector on-board the BepiColombo spacecraft. *Nuclear Instruments and Methods in Physics Research, Section A: Accelerators, Spectrometers, Detectors and Associated Equipment, 605*(3), 344–349.
25. Eisen, Y., Evans, L. G., Floyd, S., Schlemm, C., Starr, R., & Trombka, J. (2002). Radiation damage of Schottky CdTe detectors irradiated by 200 MeV protons. *Nuclear Instruments and Methods in Physics Research, Section A: Accelerators, Spectrometers, Detectors and Associated Equipment, 491*(1–2), 176–180.
26. Shvydka, D., Parsai, E. I., & Kang, J. (2008). Radiation hardness studies of CdTe thin films for clinical high-energy photon beam detectors. *Nuclear Instruments and Methods in Physics Research, Section A: Accelerators, Spectrometers, Detectors and Associated Equipment, 586*(2), 169–173.
27. Cavallini, A., et al. (2001). Irradiation-induced defects in CdTe and CdZnTe detectors. *Nuclear Instruments and Methods in Physics Research, Section A: Accelerators, Spectrometers, Detectors and Associated Equipment, 458*(1–2), 392–399.
28. Rossa, E., Schmickler, H., Brambilla, A., Verger, L., & Mongellaz, F. (2001). New development of a radiation-hard polycrystalline CdTe detector for LHC luminosity monitoring. CERN-SL-2001-024 BI.
29. Šagátová, A., et al. (2017). Radiation hardness of GaAs sensors against gamma-rays, neutrons and electrons. *Applied Surface Science, 395*, 66–71.
30. Pattabi, M., Krishnan, S., Ganesh, & Mathew, X. (2007). Effect of temperature and electron irradiation on the I–V characteristics of Au/CdTe Schottky diodes. *Solar Energy, 81*(1), 111–116.
31. Šagátová, A., Zaťko, B., Nečas, V., & Fülöp, M. (2020). Radiation hardness limits in gamma spectrometry of semi-insulating GaAs detectors irradiated by 5 MeV electrons. *Journal of Instrumentation, 15*, C01024.
32. Shiraki H. et al. (2009). THM growth and characterization of 100 mm diameter CdTe single crystals. *IEEE Transactions on Nuclear Science 56*(4), 5204617, 1717–1723.

33. Sedlačková, K., Zaťko, B., Pavlovič, M., Šagátová, A., & Nečas, V. (2020). Effects of electron irradiation on spectrometric properties of Schottky barrier CdTe radiation detectors. *International Journal of Modern Physics: Conference Series, 50*, 2060017.
34. Zaťko, B., et al. (2018). Schottky barrier detectors based on high quality 4H-SiC semiconductor: Electrical and detection properties. *Applied Surface Science, 461*(July), 276–280.
35. Frank, F. (2021). *SeGaSA software*. [Online]. Available: https://github.com/FilipFr/SeGaSA
36. Sedlačková, K., & Frank, F. (2021). Application software for automatic time- dependent spectral analysis. In *AIP Conference Proceedings, Applied Physics of Condensed Matter APCOM 2021* (pp. 080011–080011).
37. Hovington, P., Drouin, D., & Gauvin, R. (1997). CASINO: A new Monte Carlo code in C language for electron beam interaction — Part I: Description of the program. *Scanning, 19*, 1–14.
38. Lazurik, V. T., Lazurik, V. M., Popov, G., Rogov, Y., & Zimek Z. (2011). Information system and software for quality control of radiation processing. In: *IAEA: Collaborating center for radiation processing and industrial dosimetry* (p. 232).
39. Sedlačková, K., Zaťko, B., Šagátová, A., & Nečas, V. (2022). Polarization effect of Schottky-barrier CdTe semiconductor detectors after electron irradiation. *Nuclear Instruments and Methods in Physics Research A, 1027*, 166282.
40. Sedlačková, K., Zaťko, B., & Nečas, V. (In press). Spectrometry of electron irradiated CdTe Schottky-barrier semiconductor detectors before polarization onset. In *AIP conference proceedings, applied physics of condensed matter APCOM 2022*.

A Path to Produce High-Performance CdZnTe Crystals for Radiation Detection Applications: Crystal Growth by THM, Surface Preparation, and Electrode Deposition

Mustafa Ünal and Raşit Turan

1 Introduction

CdZnTe crystals have gained attention in recent years for radiation detection applications owing to their high spectral resolution and high sensitivity. Production of cadmium zinc telluride (CdZnTe) semiconductor crystals is challenging because of the native properties of the material. Low thermal conductivity, high ionicity of bonds, and segregation of Zn in the CdZnTe matrix result in multi-crystalline ingot with high defect density and non-uniformity of the composition in melt growth techniques [1–3]. Furthermore, CdZnTe is known as a soft-brittle material [4]. Thus, each mechanical process induces deep subsurface damage. Additional etching processes are necessary in order to clean the surface from damaged layers and contaminants. A clean surface is also necessary in order to obtain a metal-semiconductor interface with a low barrier and low defect density. On the other hand, contact deposition is another challenge in revealing the true potential of the crystal. The electrode deposition technique is one of the key factors determining the performance of a detector. The electroless gold plating technique is usually preferred

M. Ünal (✉)
Graduate School of Micro and Nanotechnology, Middle East Technical University, Ankara, Turkey

Crystal Growth Laboratory, MiddleEast Technical University, Ankara, Turkey
e-mail: mustafa.unal@metu.edu.tr

R. Turan
Graduate School of Micro and Nanotechnology, Middle East Technical University, Ankara, Turkey

Crystal Growth Laboratory, Middle East Technical University, Ankara, Turkey

Physics Department, Middle East Technical University, Ankara, Turkey

© The Author(s), under exclusive license to Springer Nature Switzerland AG 2023
L. Abbene, K. (Kris) Iniewski (eds.), *High-Z Materials for X-ray Detection*,
https://doi.org/10.1007/978-3-031-20955-0_12

227

for metal coating on CdZnTe surface in order to avoid adhesion problems occurring in physical vapor deposition techniques.

This chapter covers different aspects of the preparation of CdZnTe detector crystals from growth to electrode deposition. The traveling heater method is covered for the growth of CdZnTe crystals. Defects that are present in the crystal are discussed, and an attempt to summarize doping in CdZnTe is made. Electroless gold deposition and surface preparation techniques are shortly explained, and results in the literature are compared. Furthermore, the properties of THM-grown crystals are presented in different aspects.

2   Traveling Heater Method

Traveling heater method is a solution-based growth technique. It is developed in the 1950s to grow GaP from Ga solution [5–7]. This technique relies on the formation of a solution region having a lower melting point than grown material. The solution zone promotes dissolving feed material while initiating solidification at the seed surface by slowly moving the heat profile. For the growth of CdZnTe, the growth temperature is around 700–900 °C, while it is above 1100 °C for the melt growth techniques. In melt growth techniques, CdZnTe is required to be heated 20–50 °C above the melting point because of the high ionicity of bounds. This makes the usage of seed crystals impossible. On the other hand, THM enables the usage of seed crystals because growth is conducted below the melting point. Additionally, usage of seed crystal ensures obtaining single or large grains. Due to the fact that there is no complete melting of feed material, ingot composition is relatively uniform. Segregation of Zn is only observed at the end of the growth. During growth, contaminants can be trapped in the molten region, moved through the ingot, and the purity of the material can be increased. In the end, a high-quality detector CdZnTe crystal with high compositional uniformity can be obtained by this method [8–14]. Most commercial detector crystals are grown with this technique.

The solvent for CdTe/CdZnTe is Te. CdZnTe seed crystal is placed at the bottom of the growth crucible. On top of it, Te or Te-rich CdZnTe is added to form a molten/solution region. This region has a lower melting point than CdTe/CdZnTe, depending on the composition. Usage of Te-rich CdZnTe as the molten region has the advantage of limited melting of the seed crystal and reaching equilibrium in a short time. On top of the molten region, feed CdZnTe is added. A schematic of a classical THM furnace and growth material is presented in Fig. 1. The molten region is heated to growth temperature, and after it reaches equilibrium, the furnace is started to move through the feed CdZnTe material. This movement results in the heating of the feed material near the molten region-feed CdZnTe interface and causes its dissolution. On the other hand, at the seed surface, temperature decreases because heaters move away and deposition of solid CdZnTe occurs. With slow growth rates, seeded growth can be achieved. Typical reported growth rates are 2–

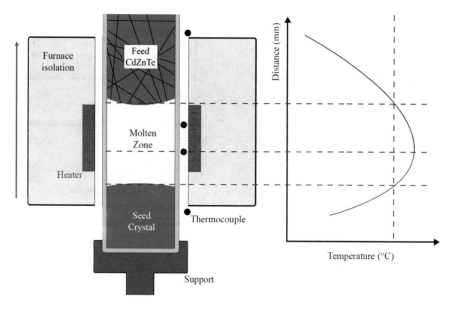

Fig. 1 The schematic of THM furnace and temperature profile

10 mm/day, while the maximum is reported as 15 mm/day, which is state of the art today [15–17].

The temperature profile is another critical key parameter for the successful THM growth of CdZnTe crystals. Temperature gradient enables mass transport from feed CdZnTe to seed CdZnTe. In Fig. 1, the temperature profile of a THM furnace is presented with corresponding crucible positions. The temperature gradient at the seed-molten region interface is reported as 45 K/cm [18]. Equilibrium temperatures at the seed-molten region interface and molten region-feed material interface should be equal [19] in order to conduct crystal growth.

3 Crystal Defects

In order to obtain high-performance detector crystals, understanding and eliminating crystal defects are important. There are several defects in the grown CdZnTe ingots, such as grain boundaries, twin boundaries, subgrain boundaries, dislocations, inclusions, etc. While the effect of grain boundary on spectral performance is deadly, as can be seen in Fig. 2a, b, twin boundaries cause slight distortions in the obtained spectra, which is shown in Fig. 2c, d. Thus, it is crucial to eliminate both defects during the cutting of the ingot. Furthermore, subgrain boundaries cannot be eliminated due to the nature of the crystal. Subgrain boundaries form a cellular network inside the crystal [20–22].

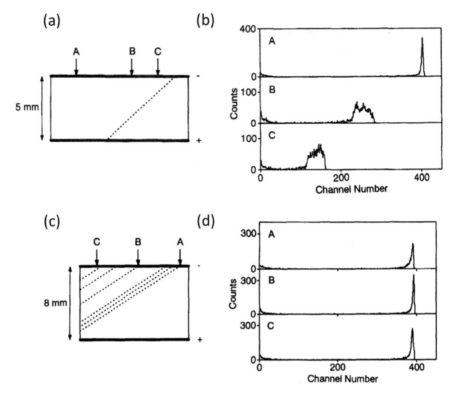

Fig. 2 The effect of grain boundary and twin boundaries on detector performance. (**a**) Illumination point on the detector for [241] Am alpha source. The dash line shows the grain boundary. (**b**) obtained [214] Am spectrums for three position points. (**c**) Illumination point on the detector for [241] Am alpha source. Dash lines show twin boundaries. (**d**) Obtained [214] Am spectrums for three position points [22]

Another performance-limiting defect is secondary phase formation. Due to distortion from stoichiometry, Te inclusions can form inside the ingot. The effect of Te inclusions on detector performance is studied in several research articles. With the increase in the size of Te inclusions, concentrations should be decreased to preserve detector performance [23].

Even though the aforementioned defects are affecting the crystal performance, major performance-limiting defects are point defects. CdZnTe and CdTe crystals are prone to have point defects. Thus, undoped crystals usually possess low resistivity values. Cd vacancies (V_{Cd}), Cd antisites (Cd_{Te}), interstitial Cd (Cd_i), Te vacancies (V_{Te}), Te antisites (Te_{Cd}), and Te interstitials (Te_i) are some primary point defects in the CdTe structure. The concentration of point defects varies with growth conditions. Furthermore, it is reported that the addition of Zn into the crystal matrix increases the solubility of Te, and intrinsic defect concentration increases [24, 25]. In Te-rich conditions, Cd vacancies and Te antisite defects are dominant in the crystal and make the crystal p-type. V_{Cd} is a shallow acceptor with the energy of 0.13 and 0.21 eV above the valance band, while CdTe is a shallow donor just below 0.1 eV from the conduction band [24, 26]. It is concluded by many researchers that shallow defects having energies minimum than 0.3 eV have no effect on the transport of charge carriers [25]. However, deep donor defects such as Te_{Cd} (0.34 eV), Cd_i (0.45 eV), Te_{Cd} (0.59 eV), V_{Te} (0.71 eV), and deep acceptor defect, Te_i (0.57 eV), are reported as performance-degrading point defects [24, 26].

In the literature, the effect of these defects on detector performance and the presence of some of these defects are still a controversial issue. Some researchers suggest increasing the concentration of deep defects for Fermi level pinning to increase resistivity [27, 28]. In contrast, other researchers propose to decrease Cd vacancies and form defect complexes in the mid bandgap for pinning of Fermi level [24, 26].

It is believed that high resistivity is obtained by Fermi level pinning near mid-bandgap. This pinning mechanism can be accomplished by using deep defect states created by dopants like Ge and Sn. However, it is reported that crystals produced by using these dopants suffer from electron trapping at the deep donor level [29]. Doping by shallow dopants like In and Cl is proved to be successful. It is observed that when doping concentration with shallow dopants is increased, resistivity increases to a certain value and stabilizes. Moreover, when doping concentration is increased further, even though resistivity stays the same, transport properties start to decline. This issue raised a question about the doping mechanism. Indium doping eliminates shallow acceptors, V_{Cd} (0.13 eV above valance band), and forms shallow donor states, In_{Cd} (0.05 eV below conduction band). On the other hand, Cl doping fills deep donor states, V_{Te} (0.71 eV below conduction band), and creates a shallow donor state, Cl_{Te} (0.35 eV below conduction band). All defect states related to doping are calculated by [110], as shown in Fig. 3. One should expect to decrease resistivity after a certain threshold value. Thus, it is speculated that doping somehow causes an increase in Te antisite defect concentration as a side effect, and Fermi level pinning is achieved. However, transport properties should be severed in this scenario.

Another approach for Fermi level pinning is a self-limiting doping mechanism. There are several complexes formed by doping. AX center is reported to be a limiting factor for p-type doping. This complex is formed by breaking two bounds and distortion in bound structure. DX centers, on the other hand, are formed by single bound breaking, These defect complexes are believed to be limiting factors for doping [26]. There is also a neutral complex formed by doping, $\left[2In_{Cd}^{+} \cdot V_{Cd}^{2-} \right]^{0}$, but this complex is not effective on transport properties [30]. Furthermore, A center complex in CdZnTe crystals is formed by doping. This complex is formed by one Cd vacancy and one substitutional In forming a $\left[In_{Cd}^{+} \cdot V_{Cd}^{2-} \right]^{-}$ complex, named as A−. On the other hand, Cl doping could form A center by a $\left[V_{Cd}^{2-} \cdot Cl_{Te}^{+} \right]^{-}$ complex. While doping shifts the Fermi level to mid-bandgap, the formation energy of Cd vacancies decreases. Thus, V_{Cd} concentration increase with doping. This

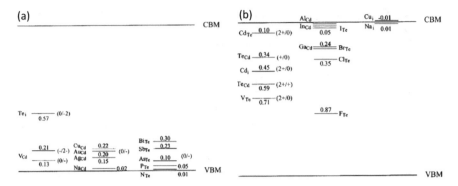

Fig. 3 Calculated (**a**) acceptor and (**b**) donor transition energy levels by [26]

phenomenon reduces the efficiency of doping. These newly formed Cd vacancies could be isolated or could form A-centers. After a certain doping level, A-centers become the dominant acceptor and limit doping efficiency. Thus, Fermi level is pinned without the need for introduction of any deep defect state. Moreover, the self-limiting mechanism enables a wide range of doping concentrations. For In doping, it is reported that doping concentrations between 10^{16} and 10^{17} cm^{-3} are acceptable, while for Cl doping, it is reported as between 10^{16} and 10^{19} cm^{-3} [26, 29, 30].

4 Surface Preparation and Electrode Deposition

The surface of $Cd_{1-x}Zn_xTe$ crystals should be lapped and polished to eliminate the damaged layer that is formed due to cutting or dicing from the ingot. This damaged layer includes an amorphous phase of $Cd_{1-x}Zn_xTe$ in addition to a cracked polycrystalline structure. Even though lapping and polishing processes remove this damaged layer from $Cd_{1-x}Zn_xTe$ crystal surface, these mechanical processes also induce a thin, damaged layer [31, 32]. This layer causes the current leakage from anode to cathode and diminishes the detector performance. In addition, contaminants like embedded abrasives, carbon, and oxygen on the surface induce interfacial defects on the $Cd_{1-x}Zn_xTe$-metal contact interface. Therefore, cleaning the surface from contaminants and the damaged layer is crucial, which can be achieved by chemical-mechanical polishing or chemical polishing/etching [33, 34]. Chemical-mechanical (CM) polishing process work in conjunction with chemical etching and mechanical surface removal processes. On the other hand, chemical polishing only depends on chemical interactions between the solution and the crystal. Usually, chemical polishing solutions can be used in the chemical-mechanical polishing process and vice versa. Even though there are several chemical polishing solutions proposed in the literature, a commonly used solution for these processes is the bromine-methanol solution [33, 35–39]. Bromine-ethylene glycol (Br:$(CH_2OH)_2$)

[39], bromine-acetic acid-ethylene glycol ($Br:C_2H_6O_3:(CH_2OH)_2$) [38], hydrogen bromide-hydrogen peroxide- ethylene glycol ($HBr:H_2O_2:(CH_2OH)_2$) [40], potassium dichromate-nitric acid-ethylene glycol ($K_2Cr_2O_7:HNO_3:(CH_2OH)_2$) [40], and hydrogen bromide-hydrogen peroxide-acetic acid ($HBr:H_2O_2:C_2H_6O_2$) [41] are some of the solutions proposed as an alternative to bromine-methanol solution. However, the usage of bromine-methanol solution eliminates complicated reactions and offers a more stable solution than other bromine-included ones. Nonetheless, bromine-methanol solution ages over time; solution temperature increases, and pH decreases due to the reaction between bromine and methanol [42]. It is essential to conduct experiments with freshly prepared solutions.

$$Cd_{1-x}Zn_xTe + Br_2 \rightarrow (1-x)CdBr_2 + x\,ZnBr_2 + Te \qquad (1)$$

The bromine-methanol solution reacts with the $Cd_{1-x}Zn_xTe$ crystal surface in such a way that cadmium dibromide ($CdBr_2$) and zinc dibromide ($ZnBr_2$) form while elemental tellurium remains on the surface as given in Eq. 1. The remaining Te is the major problem of using the bromine-methanol solution, which disturbs the stoichiometry of the surface. The remaining Te has high conductivity than the $Cd_{1-x}Zn_xTe$ crystal, and it oxidizes quickly, which results in a poor performance in the detector crystal due to the barrier at the interface of the metal-semiconductor [39]. In order to prevent these problems, etching duration should be adjusted very carefully so that stoichiometry is preserved, while the surface is cleaned from contaminants and the damaged layer.

In addition, the electrode deposition technique is one of the key factors determining the performance of a detector. Electrode metal can be coated on CdZnTe by physical vapor deposition or electroless plating techniques. Electroless plating is widely used because both surfaces can be coated simultaneously; an oxide-free interface can be obtained effortlessly, and due to the chemical reaction between solution and CdZnTe, a stronger chemical bond between metal and semiconductor can easily be achieved [43, 44]. There are several metals proposed to be used on CdZnTe crystals. Gold is a widely used metal on CdZnTe surfaces thanks to the low Schottky barrier between metal and CdZnTe. However, usage of platinum, ruthenium [43], and indium [45] is reported for several applications. Furthermore, there are studies on the use of transparent conductive oxides such as ZnO:Al [46] and indium tin oxide (ITO) [47] that are proposed as low-barrier contacts in the literature.

Electrode contacts could be ohmic or Schottky according to the electrode material. Work function (Φ) of CdZnTe is reported as 5.1 eV [46, 47] while that of gold deposited by electroless deposition technique is also reported as 5.1 eV, which forms good ohmic contacts [47, 48]. However, physical vapor deposition techniques could result in Schottky contacts with various barrier heights according to CdZnTe crystal orientation [48]. Detector performance could be significantly affected by just changing the deposition technique. In Fig. 4, the difference between electroless coated gold contacts and sputtered gold contacts can clearly be seen. The electroless coating offers low FWHM in the spectrum of ^{241}Am.

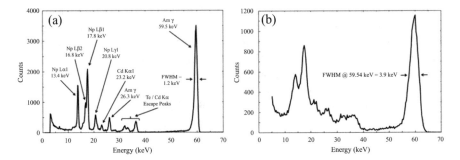

Fig. 4 The spectrum of 241Am for CdZnTe detectors having (**a**) electroless and (**b**) sputtered gold contacts [44]

Work function of platinum, ruthenium, and indium are reported as 5.12 eV, 4.71 eV, and 4.09 eV, respectively [49]. Metals having lower or higher work functions than CdZnTe are chosen to form rectifying contacts. A barrier on the interface could result in enhanced electrical properties for CdZnTe under bias voltage by limiting hole movement. There are examples of Au/CdZnTe/In structures in which gold is used for cathode and indium is used for anode [50–53].

5 CdZnTe Crystal Growth

Crystal growth experiments are conducted with the THM furnace in the Crystal Growth Laboratory of Middle East Technical University (METU-CGL). THM furnace has a 1-inch bore opening with one heating zone. Prior to growth experiments, interface shape, and position should be known. It is observed that a clean transition can be observed from seed crystal to molten zone for 700 °C furnace temperature, as shown in Fig. 5a. There is no distortion in the interface shape, and the growth starting position is observed from the cross section. For growth experiments, a multi-crystalline seed crystal is used. Seed crystal, tellurium molten zone, and feed CdZnTe material are sealed under vacuum inside carbon-coated quartz crucible. Before initiating the growth, 5 mm of the seed crystal is dissolved at 700 °C, and growth is conducted at the same temperature with a 1 mm/day transition rate. The doping of CdZnTe is conducted by 10 ppm indium. The resulting ingot is shown in Fig. 5b.

5 mm × 5 mm × 5 mm samples that are (111) orientated is cut by Logitech Diamond Wire Saw. Lapping and polishing are conducted with the Logitech PM5 polishing system. Slurry having 3 μm alumina is used for lapping purposes, while 0.3 μm alumina is used for polishing studies with polishing cloth. Chemical polishing is conducted by 5% bromine-methanol solution to remove a thin damaged and contaminated layer over the crystal. The etching process is followed by an electroless gold coating process to form planar gold electrodes. After the coating

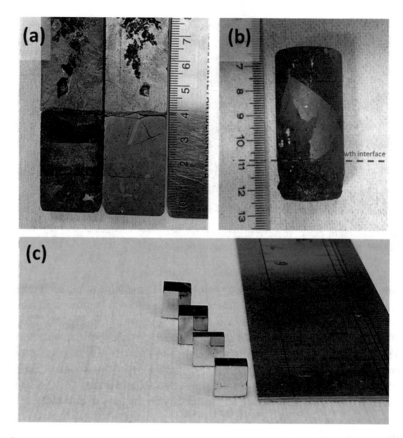

Fig. 5 (**a**) Interface position and transition region from seed to molten zone, (**b**) grown CdZnTe ingot and (**c**) CdZnTe crystals produced in METU-CGL

is completed, passivation is conducted by hydrogen peroxide (H_2O_2) solution to decrease leakage current. Prepared crystals are presented in Fig. 5c.

6 Properties of THM-Grown Crystals

Grain Evolution

The growth of CdZnTe is initiated with a multi-crystalline seed crystal. The grain structure and orientation of each grain are presented in Fig. 6a. It can be seen that the main grain is in (111) orientation, which is the closely packed plane in the zinc-blende structure. Several grains with various orientations decorate the growth interface. After 12 mm of growth (Fig. 6b), it is observed several small grains are disappeared. However, (111) oriented grain got smaller with the progress of the

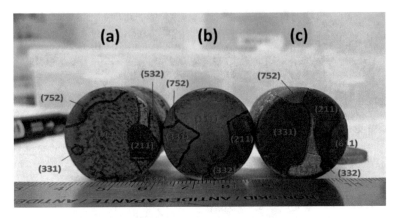

Fig. 6 Evolution of grain evolution of grown CdZnTe crystal. (**a**) beginning of the growth, (**b**) after 12 mm of the growth, and (**c**) after 23 mm of the growth

growth. It is seen that (331) oriented grain represses the large (111) oriented grain. At 23rd mm of the growth, the formation of new grains can be observed, while the surface area of (111) oriented grain is minimum, as shown in Fig. 6c. The (331) oriented grain is the dominant one in this stage of growth. It can also be seen that (211) oriented grain gets bigger. Even though the usage of (111) orientation seed is common [15, 54, 55] and growth in <111> direction is superior to others [56–58], it is speculated that the formation of new grains and suppression of (111) orientation is observed due to wall interaction and instability on the growth interface.

Structural Investigation
The structural investigation is conducted in terms of crystal quality and Te inclusion density. The crystal quality of the obtained crystal is measured by the X-ray diffraction method. Figure 7a shows the XRD measurement result from 10° to 120°. A Clear (111) peak can be seen in the measurement. Further investigation is conducted with XRD-DCRC measurement [inset of Fig. 7a]. FWHM of (333) peak is found to be 10.84 arcs. Furthermore, Te inclusion concentration is measured by IR microscope in transmission mode (Fig. 7b–e). In 5× magnification, Te inclusion concentration is calculated as 1.6×10^4 cm^{-3}. When the size distribution of Te inclusions is investigated, it can be observed that there are two separate Te inclusion peaks on the histogram given in Fig. 7c. The majority of the inclusions have diameters in the range of 10–15 µm. Te inclusions that have 8 µm diameter on average form a second peak in the histogram. The average diameter of Te inclusions in 5× magnification is found to be 11 µm, as shown in the histogram of the microscope image with a vertical red line (Fig. 7c). However, it is not easy to observe Te inclusions having diameters below 5 µm in 5× magnification. It is better to analyze Te inclusions with different magnifications. As shown in Fig. 7d, there are small inclusions that cannot be observed in 5× magnification. The majority of the inclusions have a diameter below 2 µm. The average diameter drops to 1 µm on this

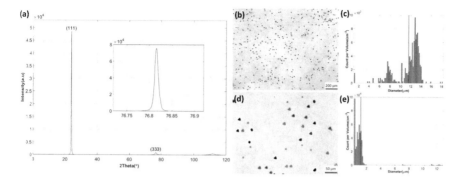

Fig. 7 (**a**) XRD measurement result for crystal obtained from growth. Inset: XRD-DCRC measurement for (333) peak. (**b**) IR microscope image of obtained crystal 5× magnification in EFI mode with Te inclusion density calculated as 1.06×10^4 cm^{-3}, (**c**) histogram of the size of Te inclusions for (**b**). (**d**) IR microscope image at 20× magnification in EFI mode with Te inclusion density calculated as 4.70×10^5 cm^{-3}, (**e**) histogram of the size of Te inclusions for (**d**). Average diameters for both (**c**) and (**e**) are marked with a straight red line in the related graphs

scale, as can be seen in Fig. 7e with the vertical red line. Te inclusion concentration is calculated as 4.70×10^5 cm^{-3} for this magnification. Even though Te inclusion concentration is increased at higher magnification, the diameter of Te inclusions is significantly decreased. In the literature, the correlation between Te inclusion size, concentration, and mobility-lifetime product of electrons is presented [59–61]. It is possible to speculate that Te inclusions observed in the prepared crystal have minimal effect on mobility-lifetime product and performance of the crystal thanks to low concentration of inclusions for different sizes.

Electrical and Spectral Characterization

Figure 8 shows electrical measurement results of CdZnTe crystal grown in METU-CGL. I-V measurement shows symmetrical behavior of the crystal between -1000 V and 1000 V (Fig. 8a). It can be seen that crystal manifests leakage current below 5 nA in this measurement range. To be able to calculate resistivity, I-V measurement is taken in -1 V to 1 V range, as shown in the inset of Fig. 8a. The resistivity of the crystal is calculated as 6.1×10^9 Ω.cm. On the other hand, the mobility-lifetime product of electrons in the CdZnTe crystal grown in METU-CGL is calculated by fitting on obtained spectrum data at different bias voltages by using the Hecht equation up to 1000 V presented in Fig. 8b. Mobility-lifetime product for electrons is calculated as 6.7×10^{-3} cm^2/V. These values are comparable with commercially available crystals (Table 1). It can be seen that lower leakage current and higher resistivity are achieved in the grown crystal.

Gamma spectrums are obtained with different radiation sources. The FWHM of the peak that appeared at 59 keV for ^{241}Am is calculated as 6.45 keV. In addition, X-ray peaks and Cd & Te escape peaks can easily be identified, as shown in Fig. 9a. On the other hand, for high energies, parallel electrode geometry is not

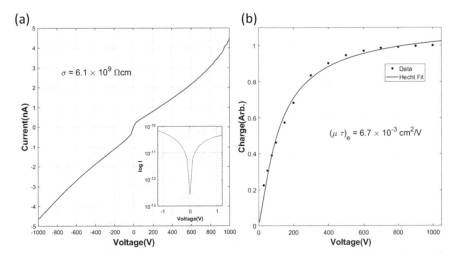

Fig. 8 (**a**) I–V measurement from −1000 V to 1000 V for CdZnTe. Inset shows I–V measurement in logarithmic scale from −1 V to 1 V. (**b**) Graph of collected charges at different bias voltages, fitted by Hecht equation

Table 1 Comparison of the electrical performance of crystal grown in METU-CGL with commercial samples

Sample	$R(\Omega)$	$\delta(\Omega.cm)$	Leakage current at 200 V	Mobility lifetime ($\frac{cm^2}{V}$)
Commercial-1	9.80×10^9	4.90×10^9	10 nA	7.6×10^{-3}
Commercial-2	1.09×10^9	2.18×10^8	25 nA	5.1×10^{-3}
METU-THM	1.22×10^{10}	6.10×10^9	2 nA	6.7×10^{-3}

effective because of the two orders of magnitude difference in mobility-lifetime product of electrons and holes. In Fig. 9b, the spectrum of ^{57}Co taken with parallel electrode geometry is presented. Even though the crystal suffers from mobility-lifetime difference, two gamma peaks can be identified alongside X-ray peaks. Furthermore, ^{137}Cs spectrum is obtained, and the gamma emission peak at 662 keV is identified with a strong background (Fig. 9c). X-ray peaks are dominant for this spectrum, shown as an inset in Fig. 9c.

7 Conclusion

In this chapter, a short review of the traveling heater method (THM), crystal defects, surface processes, and electrode deposition is made. While THM offers seeded growth with low segregation and high yield, crystal growth durations are significantly longer than melt growth techniques. It is seen that avoiding crystal defects such as grain boundaries and twin boundaries is crucial, while Te inclusion

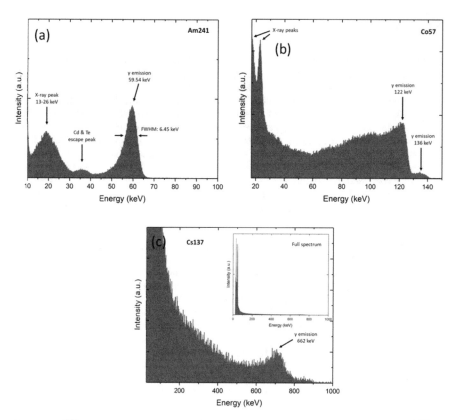

Fig. 9 (**a**) [241]Am spectrum taken from CdZnTe with parallel electrodes at 500 V, (**b**) [57]Co spectrum taken from CdZnTe with parallel electrodes at 500 V and (**c**) ^{137}Cs spectrum taken from CdZnTe with parallel electrodes at 500 V

concentration should be decreased in order to obtain high-performance detector crystals. On the other hand, major performance-limiting defects are point defects. The doping mechanism is discussed, and an attempt to explain the defect mechanism is made. Furthermore, the importance of a clean and defect-free surface is covered, which is followed by electrode deposition techniques.

CdZnTe ingot is grown in METU Crystal Growth Laboratory. It is seen that even though the large single grain is presented at the beginning of the growth, the formation of new grains occurs, and the large (111) grain gets smaller with the progress of the growth due to crystal-wall interaction at the interface. A crystal having a low density of Te inclusions is obtained with a 10.84 arc.s FWHM in XRD-DCRC measurement. It is seen that the grown crystal has 6.1×10^9 Ω.cm resistivity with below 5 nA leakage current at 1000 V bias voltage. Moreover, the mobility-lifetime product for electrons is calculated as 6.7×10^{-3} cm^{-2}/V. CdZnTe detector crystal with a comparable mobility-lifetime product and superior resistivity compared with commercial crystals is grown and processed in METU-CGL.

References

1. Shalvoy, R. B., Fisher, G. B., & Stiles, P. J. (1977). Bond ionicity and structural stability of some average-valence-five materials studied by x-ray photoemission. *Physical Review B, 15*, 1680–1697. https://doi.org/10.1103/PhysRevB.15.1680
2. Marchini, L., Zambelli, N., Piacentini, G., Zha, M., Calestani, D., Belas, E., & Zappettini, A. (2011). Characterization of CZT crystals grown by the boron oxide encapsulated vertical Bridgman technique for the preparation of X-ray imaging detectors. *Nuclear Instruments and Methods in Physics Research A, 633*, S92–S94. https://doi.org/10.1016/j.nima.2010.06.133
3. Datta, A., Swain, S., Bhaladhare, S., & Lynn, K. G. (2012). Experimental studies on control of growth interface in MVB grown CdZnTe and its consequences. In *IEEE nuclear science symposium conference record* (pp. 4720–4726). https://doi.org/10.1109/NSSMIC.2011.6154703
4. Zhang, Z., Guo, D., Kang, R., Gao, H., Jin, Z., & Meng, Y. (2010). Subsurface crystal lattice deformation machined by ultraprecision grinding of soft-brittle CdZnTe crystals. *International Journal of Advanced Manufacturing Technology, 47*, 1065–1081. https://doi.org/10.1007/s00170-009-2253-y
5. Wolff, G. A., Hebert, R. A., & Broder, J. D. (1955). Electroluminescence of GaP. *Physical Review, 100*, 1144–1145. https://doi.org/10.1103/PhysRev.100.1144
6. Schulze, R. G., & Petersen, P. E. (1974). Photoconductivity in solution-grown copper-doped GaP. *Journal of Applied Physics, 45*, 5307–5311. https://doi.org/10.1063/1.1663235
7. Dost, S. (2006). *Single crystal growth of semiconductors from metallic solutions*. Elsevier.
8. Roy, U. N., Gueorguiev, A., Weiller, S., & Stein, J. (2009). Growth of spectroscopic grade Cd0.9Zn0.1Te:In by THM technique. *Journal of Crystal Growth, 312*, 33–36. https://doi.org/10.1016/j.jcrysgro.2009.09.035
9. Chen, H., Awadalla, S. A., Redden, R., Bindley, G., Bolotnikov, A. E., Camarda, G. S., Carini, G., & James, R. B. (2010). High-performance, large-volume THM CdZnTe detectors for medical imaging and homeland security applications. In *IEEE nuclear science symposium conference record. Nuclear science symposium* (pp. 3629–3637). https://doi.org/10.1109/nssmic.2006.353781
10. Awadalla, S. A., Mackenzie, J., Chen, H., Redden, B., Bindley, G., Duff, M. C., Burger, A., Groza, M., Buliga, V., Bradley, J. P., Dai, Z. R., Teslich, N., & Black, D. R. (2010). Characterization of detector-grade CdZnTe crystals grown by traveling heater method (THM). *Journal of Crystal Growth, 312*, 507–513. https://doi.org/10.1016/j.jcrysgro.2009.11.007
11. Peterson, J. H. (2017). *Understanding growth rate limitations in production of single-crystal cadmium zinc telluride (CZT) by the traveling heater method (THM)*. The University of Minnesota.
12. Ghaddar, C. K., Lee, C. K., Motakef, S., & Gillies, D. C. (1999). Numerical simulation of THM growth of CdTe in presence of rotating magnetic fields (RMF). *Journal of Crystal Growth, 205*, 97–111. https://doi.org/10.1016/S0022-0248(99)00206-7
13. Roy, U. N., Weiler, S., Stein, J., Cui, Y., Groza, M., Buliga, V., & Burger, A. (2012). Zinc mapping in THM grown detector grade CZT. *Journal of Crystal Growth, 347*, 53–55. https://doi.org/10.1016/j.jcrysgro.2012.03.013
14. Roy, U. N., Weiler, S., & Stein, J. (2010). Growth and interface study of 2 in diameter CdZnTe by THM technique. *Journal of Crystal Growth, 312*, 2840–2845. https://doi.org/10.1016/j.jcrysgro.2010.05.046
15. Shiraki, H., Funaki, M., Ando, Y., Kominami, S., Amemiya, K., & Ohno, R. (2010). Improvement of the productivity in the THM growth of CdTe single crystal as nuclear radiation detector. *IEEE Transactions on Nuclear Science, 57*, 395–399. https://doi.org/10.1109/TNS.2009.2035316
16. Shiraki, H., Funaki, M., Ando, Y., Tachibana, A., Kominami, S., & Ohno, R. (2009). THM growth and characterization of 100 mm diameter CdTe single crystals. *IEEE Transactions on Nuclear Science, 56*, 1717–1723. https://doi.org/10.1109/TNS.2009.2016843

17. Hiroyuki, S., Funaki, M., Ando, Y., Kominami, S., Amemiya, K., & Ohno, R. (2007). Improvement of the productivity in the growth of CdTe single crystal by THM for the new PET system. In *IEEE nuclear science symposium conference record* (pp. 1783–1787).
18. Zhou, B., Jie, W., Wang, T., Xu, Y., Yang, F., Yin, L., Zhang, B., & Nan, R. (2018). Growth and characterization of detector-grade Cd0.9Zn0.1Te crystals by the traveling heater method with the accelerated crucible rotation technique. *Journal of Electronic Materials, 47*, 1125–1130. https://doi.org/10.1007/s11664-017-5853-6
19. Hong, B., Zhang, S., Zheng, L., Zhang, H., Wang, C., & Zhao, B. (2020). Controlling nucleation during unseeded THM growth of CdZnTe crystal. *Journal of Crystal Growth., 534*, 125482. https://doi.org/10.1016/j.jcrysgro.2020.125482
20. Roy, U. N., Camarda, G. S., Cui, Y., Gul, R., Yang, G., Zazvorka, J., Dedic, V., Franc, J., & James, R. B. (2019). Evaluation of CdZnTeSe as a high-quality gamma-ray spectroscopic material with better compositional homogeneity and reduced defects. *Scientific Reports, 9*. https://doi.org/10.1038/s41598-019-43778-3
21. Sajjad, M., Chaudhuri, S. K., Kleppinger, J. W., Karadavut, O., & Mandal, K. C. (2020). Investigation on Cd0.9Zn0.1Te1-ySey single crystals grown by vertical Bridgman technique for high-energy gamma radiation detectors. In *SPIE-Intl Soc Optical Eng* (p. 49). https://doi.org/10.1117/12.2570592
22. Luke, P. N., & Eissler, E. E. (1996). Performance of CdZnTe coplanar-grid gamma-ray detectors. *IEEE Transactions on Nuclear Science, 43*, 1481–1486. https://doi.org/10.1109/23.507088
23. Bolotnikov, A. E., Camarda, G. S., Carini, G. A., Cui, Y., Kohman, K. T., Li, L., Salomon, M. B., & James, R. B. (2007). Performance-limiting defects in CdZnTe detectors. *IEEE Transactions on Nuclear Science, 54*, 821–827. https://doi.org/10.1109/TNS.2007.894555
24. Szeles, C. (2004). Advances in the crystal growth and device fabrication technology of CdZnTe room temperature radiation detectors. *IEEE Transactions on Nuclear Science, 51*, 1242–1249.
25. Szeles, C. (2004). CdZnTe and CdTe materials for X-ray and gamma ray radiation detector applications. *Physica Status Solidi (B) Basic Research, 241*, 783–790. https://doi.org/10.1002/pssb.200304296
26. Wei, S. H., & Zhang, S. B. (2002). Chemical trends of defect formation and doping limit in II–VI semiconductors: The case of CdTe. *Physical Review B - Condensed Matter and Materials Physics, 66*, 1–10. https://doi.org/10.1103/PhysRevB.66.155211
27. Fiederle, M., Fauler, A., Babentsov, V., Konrath, J. P., & Franc, J. (2004). Growth of high-resistivity CdTe and (Cd,Zn)Te crystals. *Hard X-Ray and Gamma-Ray Detector Physics V, 5198*, 48. https://doi.org/10.1117/12.506032
28. Fiederle, M., Fauler, A., Konrath, J., Babentsov, V., Franc, J., & James, R. B. (2004). Comparison of undoped and doped high resistivity CdTe and (Cd,Zn)Te detector crystals. *IEEE Transactions on Nuclear Science, 51*, 1864–1868. https://doi.org/10.1109/TNS.2004.832958
29. Biswas, K., & Du, M. H. (2012). What causes high resistivity in CdTe. *New Journal of Physics, 14*. https://doi.org/10.1088/1367-2630/14/6/063020
30. Li, Q., Jie, W., Fu, L., Wang, T., Yang, G., Bai, X., & Zha, G. (2006). Optical and electrical properties of indium-doped Cd 0.9 Zn 0.1 Te crystal. *Journal of Crystal Growth, 295*, 124–128. https://doi.org/10.1016/j.jcrysgro.2006.07.030
31. Zhang, Z., Gao, H., Jie, W., Guo, D., Kang, R., & Li, Y. (2008). Chemical mechanical polishing and nanomechanics of semiconductor CdZnTe single crystals. *Semiconductor Science and Technology, 23*. https://doi.org/10.1088/0268-1242/23/10/105023
32. Zhang, Z., Meng, Y., Guo, D., Kang, R., & Gao, H. (2010). Nanoscale machinability and subsurface damage machined by CMP of soft-brittle CdZnTe crystals. *International Journal of Advanced Manufacturing Technology, 47*, 1105–1112. https://doi.org/10.1007/s00170-009-2225-2

33. Rouse, A. A., Szeles, C., Ndap, J. O., Soldner, S. A., Parnham, K. B., Gaspar, D. J., Engelhard, M. H., Lea, A. S., Shutthanandan, S. V., Thevuthasan, T. S., & Baer, D. R. (2002). Interfacial chemistry and the performance of bromine-etched CdZnTe radiation detector devices. *IEEE Transactions on Nuclear Science, 49*(I), 2005–2009. https://doi.org/10.1109/TNS.2002.801705

34. Bensouici, A., Carcelen, V., Plaza, J. L., De Dios, S., Vijayan, N., Crocco, J., Bensalah, H., Dieguez, E., & Elaatmani, M. (2010). Study of effects of polishing and etching processes on Cd 1-xZnxTe surface quality. *Journal of Crystal Growth.* https://doi.org/10.1016/j.jcrysgro.2010.03.045

35. Özsan, M. E., Sellin, P. J., Veeramani, P., Hinder, S. J., Monnier, M. L. T., Prekas, G., Lohstroh, A., & Baker, M. A. (2010). Chemical etching and surface oxidation studies of cadmium zinc telluride radiation detectors. *Surface and Interface Analysis, 42*, 795–798. https://doi.org/10.1002/sia.3146

36. Babar, S., Sellin, P. J., Watts, J. F., & Baker, M. A. (2013). An XPS study of bromine in methanol etching and hydrogen peroxide passivation treatments for cadmium zinc telluride radiation detectors. *Applied Surface Science, 264*, 681. Faculty of Engineering and Physical Sciences, University of.

37. Bensalah, H., Plaza, J. L., Crocco, J., Zheng, Q., Carcelen, V., Bensouici, A., & Dieguez, E. (2011). The effect of etching time on the CdZnTe surface. *Applied Surface Science, 257*, 4633–4636. https://doi.org/10.1016/j.apsusc.2010.12.103

38. Mescher, M. J., James, R. B., Schlesinger T. E., & Hermon, H. (2000). Method for surface passivation and protection of cadmium zinc telluride crystals, US006043106A. https://doi.org/10.1016/0375-6505(85)90011-2.

39. Zázvorka, J., Franc, J., Beran, L., Moravec, P., Pekárek, J., & Veis, M. (2016). Dynamics of native oxide growth on CdTe and CdZnTe X-ray and gamma-ray detectors. *Science and Technology of Advanced Materials, 17*, 792–798. https://doi.org/10.1080/14686996.2016.1250105

40. Hossain, A., Bolotnikov, A. E., Camarda, G. S., Cui, Y., Jones, D., Hall, J., Kim, K. H., Mwathi, J., Tong, X., Yang, G., & James, R. B. (2014). Novel approach to surface processing for improving the efficiency of CdZnTe detectors. *Journal of Electronic Materials, 43*, 2771–2777. https://doi.org/10.1007/s11664-013-2698-5

41. Okwechime, I. O., Egarievwe, S. U., Hossain, A., Hales, Z. M., Egarievwe, A. A., & James, R. B. (2014). Chemical treatment of CdZnTe radiation detectors using hydrogen bromide and ammonium-based solutions. *Hard X-Ray, Gamma-Ray, and Neutron Detector Physics XVI, 9213*, 165–169. https://doi.org/10.1117/12.2063067

42. Bowman, P. T., Ko, E. I., & Sides, P. J. (1990). A potential hazard in preparing bromine-methanol solutions. *Journal of the Electrochemical Society, 137*, 1309–1311. https://doi.org/10.1149/1.2086655

43. Zheng, Q., Dierre, F., Corregidor, V., Crocco, J., Bensalah, H., Plaza, J. L., Alves, E., & Dieguez, E. (2012). Electroless deposition of Au, Pt, or Ru metallic layers on CdZnTe. *Thin Solid Films, 525*, 56–63. https://doi.org/10.1016/j.tsf.2012.09.058

44. Bell, S. J., Baker, M. A., Duarte, D. D., Schneider, A., Seller, P., Sellin, P. J., Veale, M. C., & Wilson, M. D. (2017). Performance comparison of small-pixel CdZnTe radiation detectors with gold contacts formed by sputter and electroless deposition. *Journal of Instrumentation, 12*. https://doi.org/10.1088/1748-0221/12/06/P06015

45. Kim, K. H., Cho, S. H., Suh, J. H., Won, J. H., Hong, J. K., & Kim, S. U. (2009). Schottky-type polycrystalline CdZnTe X-ray detectors. *Current Applied Physics, 9*, 306–310. https://doi.org/10.1016/j.cap.2008.01.020

46. Roy, U. N., Camarda, G. S., Cui, Y., Gul, R., Hossain, A., Yang, G., Mundle, R. M., Pradhan, A. K., & James, R. B. (2017). Assessment of a new ZnO:Al contact to CdZnTe for X- and gamma-ray detector applications. *AIP Advances, 7*, 1–6. https://doi.org/10.1063/1.5001701

47. Li, L., Xu, Y., Zhang, B., Wang, A., Dong, J., Yu, H., & Jie, W. (2018). Preparation of indium tin oxide contact to n-CdZnTe gamma-ray detector. *Applied Physics Letters, 112*. https://doi.org/10.1063/1.5023133

48. Zha, G., Jie, W., Tan, T., Zhang, W., & Xu, F. (2007). The interface reaction and schottky barrier between metals and CdZnTe. *Journal of Physical Chemistry C, 111*, 12834–12838. https://doi.org/10.1021/jp0734070

49. Hölzl, J., & Schulte, F. K. (1979). Work function of metals. In J. Hölzl, F. K. Schulte, & H. Wagner (Eds.), *Solid surface physics* (pp. 1–150). Springer. https://doi.org/10.1007/BFb0048919

50. Chen, H., Awadalla, S. A., Marthandam, P., Iniewski, K., Lu, P. H., & Bindley, G. (2009). CZT device with improved sensitivity for medical imaging and homeland security applications. *Hard X-Ray, Gamma-Ray, and Neutron Detector Physics XI, 7449*, 15–31. https://doi.org/10.1117/12.828514

51. Zázvorka, J., Franc, J., Dědič, V., & Hakl, M. (2014). Electric field response to infrared illumination in CdTe/CdZnTe detectors. *Journal of Instrumentation, 9*. https://doi.org/10.1088/1748-0221/9/04/C04038

52. Narita, T., Bloser, P. F., Grindlay, J. E., & Jenkins, J. A. (2000). Development of gold-contacted flip-chip detectors with IMARAD CZT. *Hard X-Ray, Gamma-Ray, and Neutron Detector Physics II, 4141*, 89–96. https://doi.org/10.1117/12.407569

53. Narita, T., Grindlay, J. E., Jenkins, J. A., Perrin, M., Marrone, D., Murray, R., & Connell, B. (2002). Design and preliminary tests of a prototype CZT imaging array. *X-Ray and Gamma-Ray Instrumentation for Astronomy XII, 4497*, 79–87. https://doi.org/10.1117/12.454234

54. Funaki, M., Ozaki, T., Satoh, K., & Ohno, R. (1999). Growth and characterization of CdTe single crystals for radiation detectors. *Nuclear Instruments and Methods in Physics Research, Section A: Accelerators, Spectrometers, Detectors and Associated Equipment, 436*, 120–126. https://doi.org/10.1016/S0168-9002(99)00607-5

55. Xu, Y., Jie, W., Sellin, P. J., Wang, T., Fu, L., Zha, G., & Veeramani, P. (2009). Characterization of CdZnTe crystals grown using a seeded modified vertical Bridgman method. *IEEE Transactions on Nuclear Science, 56*, 2808–2813. https://doi.org/10.1109/TNS.2009.2026277

56. Asahi, T., Oda, O., Taniguchi, Y., & Koyama, A. (1996). Growth and characterization of 100 mm diameter CdZnTe single crystals by the vertical gradient freezing method. *Journal of Crystal Growth, 161*, 20–27. https://doi.org/10.1016/0022-0248(95)00606-0

57. Hassani, S., Lusson, A., Tromson-Carli, A., & Triboulet, R. (2003). Seed-free growth of (1 1 1) oriented CdTe and CdZnTe crystals by solid-state recrystallization. *Journal of Crystal Growth, 249*, 121–127. https://doi.org/10.1016/S0022-0248(02)02114-0

58. Ivanov, Y. M. (1998). The growth of single crystals by the self-seeding technique. *Journal of Crystal Growth, 194*, 309–316. https://doi.org/10.1016/S0022-0248(98)00620-4

59. Mao, Y., Zhang, J., Min, J., Liang, X., Huang, J., Tang, K., Ling, L., Li, M., Zhang, Y., & Wang, L. (2018). Study of Te inclusion and related point defects in THM-growth CdMnTe crystal. *Journal of Electronic Materials, 47*, 4239–4248. https://doi.org/10.1007/s11664-018-6117-9

60. Bolotnikov, A. E., Babalola, S., Camarda, G. S., Cui, Y., Gul, R., Egarievwe, S. U., Fochuk, P. M., Fuerstnau, M., Horace, J., Hossain, A., Jones, F., Kim, K. H., Kopach, O. V., McCall, B., Marchini, L., Raghothamachar, B., Taggart, R., Yang, G., Xu, L., & James, R. B. (2011). Correlations between crystal defects and performance of CdZnTe detectors. *IEEE Transactions on Nuclear Science, 58*, 1972–1980. https://doi.org/10.1109/TNS.2011.2160283

61. Bolotnikov, A. E., Camarda, G. S., Carini, G. A., Cui, Y., Li, L., & James, R. B. (2007). Cumulative effects of Te precipitates in CdZnTe radiation detectors. *Nuclear Instruments and Methods in Physics Research, Section A: Accelerators, Spectrometers, Detectors and Associated Equipment, 571*, 687–698. https://doi.org/10.1016/j.nima.2006.11.023

Index

© The Author(s), under exclusive license to Springer Nature Switzerland AG 2023
L. Abbene, K. (Kris) Iniewski (eds.), *High-Z Materials for X-ray Detection*,
https://doi.org/10.1007/978-3-031-20955-0

Printed in the United States
by Baker & Taylor Publisher Services